NELSON MATHS

AUSTRALIAN CURRICULUM NSW

4

Teacher's Resource Book

Pauline Rogers

Contents

BLM INDEX (SEE REPRODUCIBLE RESOURCES)

NELSON TEACHING OBJECTS (NTOS) INDEX

Introduction

Nelson Maths: Australian Curriculum NSW supports the Australian Curriculum content strands *Number and Algebra, Measurement and Geometry* and *Statistics and Probability* F–6, and integrates the proficiency strands of *Understanding, Fluency, Problem Solving* and *Reasoning* throughout the activities and tasks.

This edition of *Nelson Maths* continues the tradition of providing teachers with choice by featuring an array of hands-on tasks and investigative activities. Teachers have the opportunity to tailor a program for their students based on their different learning styles and diverse needs. Each unit provides ideas on how to scaffold students' learning and also provides learning tasks to extend more able students.

Assessment is key in ***Nelson Maths: Australian Curriculum NSW*** – teachers are provided with a variety of opportunities to assess their students' learning for future planning. Teachers can choose assessment methods to best suit individual students, providing them with the opportunity to demonstrate what they have learned. The program provides specific recommendations for future learning experiences, which are linked to the assessment tasks.

Nelson Maths: Australian Curriculum NSW:

- enables teachers to implement focused teaching
- assists students by scaffolding their learning
- improves students' mathematical understandings and thinking
- provides open-ended and stimulating tasks, allowing students to work at their appropriate developmental level
- caters for various and individual learning styles
- develops students' mental computation skills
- provides teachers with choices so they can readily meet the needs of individual students and/or groups of students
- allows for effective grouping of students
- integrates Information and Communication Technology (ICT)
- provides ongoing assessment and opportunities for a variety of assessment types.

Nelson Maths: Australian Curriculum NSW provides between 30 and 33 units of work for each year level. Each unit of work is divided into three Lesson Plans. A Lesson Plan can be taught over one or more teaching sessions, depending on the needs of students. While the Lessons Plans are sequential and should be taught in the order in which they appear in the unit, teachers have the flexibility to complete several activities from a Lesson Plan over a number of days to cater for the learning requirements of the students in their class.

ACRONYMS USED IN *NELSON MATHS: AUSTRALIAN CURRICULUM NSW*

ATC: Assessment Task Card

BLM: Blackline Master

LO: Learning Object (used in Independent Tasks)

ML: Maths Language

NTO: Nelson Teaching Object (able to be used on an interactive whiteboard or personal computer)

PDF: Portable Document Format

Components

TEACHER'S RESOURCE BOOK

The Teacher's Resource Book provides between 30 and 33 units covering the *Australian Curriculum: Mathematics* content strands of *Number and Algebra*, *Measurement and Geometry* and *Statistics and Probability*.

Each Teacher's Resource Book also provides:

- Assessment Task Cards (with linking Targeted Assessment Task Cards)
- Mid- and End-of-Year Tests (Tests A and B)
- Assessment and Planning BLMs
- Nelson Teaching Objects (NTOs), accessible via a web link
- BLMs, Tests, Answers (Year 3 and up) and Assessment Task Cards as printable PDFs , accessible via the Nelson Primary website.

Each unit is divided into three Lesson Plans. Each Lesson Plan features the following:

At the beginning of each unit the following are listed:
the link to the *Australian Curriculum: Mathematics* content strand NSW *Mathematics K–10 Syllabus* substrand and outcome for that unit.

Maths Language (ML):
In each unit is listed the vocabulary the teacher models and uses in his or her teaching, and the language students are encouraged to use.

Tuning In
This activity orientates students to the focus of the lesson and revises the skills that they will need to use in the lesson. It may provide a link to the previous lesson or previous units. It may also be used as a pre-assessment activity.

Odd and Even Numbers

Number and Algebra
Whole numbers MA2-4NA applies place value to order, read and represent numbers of up to five digits
Addition and subtraction MA2-5NA uses mental and written strategies for addition and subtraction involving two-, three-, four- and five-digit numbers
Multiplication and division MA2-6NA uses mental and informal written strategies for multiplication and division

ML addition, counting sequence, division, even, integers, multiplication, numbers, odd, pattern, subtraction, Venn diagram

LESSON PLAN 1

TUNING IN
COUNTING PATTERNS
You will need: NTO 4.1 'Hundred Chart' or an enlarged copy of BLM 1 'Hundred Chart'
Display NTO 4.1 'Hundred Chart' on the IWB. Give a student a starting number and a counting sequence. Have the student, with the help of the class, complete the counting sequence, either by selecting squares or pointing to numbers on the chart. Have the class count the sequence aloud. Ask questions, such as, 'What is the starting number of the counting sequence? What are we counting by? What patterns can we see with the numbers?' Explore counting sequences, such as by 2s, 3s and 5s, with different starting points.

WHOLE-CLASS INTRODUCTION
EXPLORING ODD AND EVEN NUMBERS
You will need: an IWB blank screen or poster paper, whiteboard markers
On the IWB or poster paper, draw a Venn diagram with the headings 'Odd Numbers' and 'Even Numbers' for each circle. As a class, have students brainstorm and record what they know about odd and even numbers in the relevant circles and the aspects that are common. This may include descriptions, examples of numbers, where these numbers may be found, and why odd and even numbers are important.

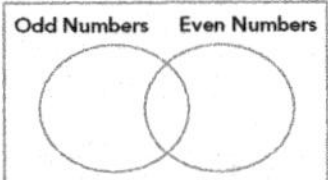

INDEPENDENT TASKS
Note: Choose from Tasks 1, 2 or 3.
You will need: long rolls of paper, LO: *L1064 'Musical number patterns: odds and evens'*, Student Book p. 4: 'Identifying Odd and Even Numbers'

TASK 1: CONTINUE THE SEQUENCE
Provide each student with a long roll of paper. Give students an even starting number, differentiated according to student needs. Have students write the starting number on the roll of paper and then continue writing the next 10 to 20 even numbers. Repeat this activity for odd numbers. Then have students form pairs to compare their sequences of even and odd numbers, looking for similarities and differences.

TASK 2: INTERACTIVE TASK
Have students work independently on computers on LO: *L1064 'Musical number patterns: odds and evens'*, completing a counting rule that matches a pattern on a number line, based on odd or even numbers.

TASK 3: STUDENT BOOK p. 4 ***'Identifying Odd and Even Numbers'***

TEACHING GROUP
You will need: large sheets of paper with two separate large circles drawn on them, counters, accessto the internet
DIVIDING COUNTERS
- For students who require support, provide large sheets of paper with two separate large circles drawn on them. Give each student a handful of counters, and ask them to divide the counters evenly into the two circles. Explore with students that if there is an even number of counters, no counters will be left over, but if there is an odd number of counters, one counter will be left over. Give students more counters and ask them to count and record the total number of counters, then predict if the number is odd or even. Then have them place the counters in the circles and clarify their answer. Repeat a number of times.

Nelson Maths Australian Curriculum NSW 20 Teacher's Resource Book Year **4**

ODD AND EVEN NUMBER EXPLORATION
- For students who require a challenge, provide them with a number set, e.g. decimals or negative numbers, and have them use the internet to find out if these numbers are odd or even. Have students collect information to share with the rest of the class – this may be in the form of a *Word* document or links that could be shared on the IWB.

REFLECTION
Select from the following to suit your class and their learning outcomes:
- Return to the Venn diagram created in the Whole-Class Introduction, and review and add new ideas about odd and even numbers. Display the chart in the classroom.
- Have students share the number sequences they made in 'Continue the Sequence', Independent Tasks, Task 1. Invite students to share what they found when they made the comparison with their partner.
- Invite students in the Teaching Group who completed the internet research to share what they found. This could be added to the Venn diagram of odd and even numbers.

LESSON PLAN 2

TUNING IN
TWO DICE
You will need: two large dice or NTO 4.2 'Dice', a large sheet of paper
Have students sit in a circle and take it in turns to roll the two dice and add the numbers together. Once students understand the activity, ask them to note whether the numbers rolled are odd or even and if they add to an odd or even number. The information could be recorded in a table on the board or on a large sheet of paper. At the end of the activity, have students draw conclusions, e.g. odd + odd = even.

WHOLE-CLASS INTRODUCTION
ADDING ODDS AND EVENS
You will need: cards with 2-digit numbers (these can be created from BLM 1 'Hundred Chart' or from pieces of card), calculators or NTO 4.3 'Calculator'
Give each student a different 2-digit number – distribute both odd and even numbers. Ask students to use their numbers to create an even number, with addition or subtraction, by finding a partner. Have them check their answer, either with pen and paper or a calculator (or NTO 4.3 'Calculator'). Equations could be recorded on the board as a group. Then have students pair up to create odd numbers, again recording equations on the board as a group. Challenge students by having them form groups of three to create odd and even numbers. Link this activity to the chart that students developed in the Tuning In activity. Ask, 'Do the rules we came up with work for 2-digit numbers?' Add any conclusions regarding subtraction to the chart, e.g. odd – even = odd.

INDEPENDENT TASKS
Note: Choose from Tasks 1, 2 or 3.
You will need: BLM 2 'Odd and Even Equations', *Excel*, Student Book p. 5 'Adding and Subtracting Odd and Even Numbers'

TASK 1: ODD AND EVEN EQUATIONS
Have students complete BLM 2 'Odd and Even Equations'.

TASK 2: INTERACTIVE TASK
Have students create an even counting pattern of five numbers down a column in *Excel*. They can then extend the pattern by selecting their five numbers and dragging the bottom right corner of the selected box down the page to continue the number pattern. Have students explore different patterns. Challenge students to explore patterns such as odd, even, odd, even.

TASK 3: STUDENT BOOK p. 5 ***'Adding and Subtracting Odd and Even Numbers'***

TEACHING GROUP
You will need: A4 sheets of paper, rulers, counters, four different-coloured dice, calculators
ADDITION AND SUBTRACTION WITH COUNTERS
- For students who require support, give them a sheet of A4 paper. Have them divide the paper into quarters and label as shown right.

odd and odd	odd and even
even and odd	even and even

Unit **1** Odd and Even Numbers 21

Whole-Class Introduction
The teacher introduces the main focus of the lesson to the whole class. Through carefully planned activities and questioning, students are introduced to new skills and understandings, which they will develop further throughout the other components of the lesson. Students who need further support or extension in a Teaching Group can be identified at this time.

Independent Tasks
Students work in small groups, pairs or individually to further strengthen their understandings and skills. The teacher may choose to have students complete an open-ended task, an interactive Learning Object, NTO or ICT task and/or the accompanying Student Book page. Depending on the learning needs of students, one or more of these tasks may be chosen.

Reflection

The class regroups as a whole to reflect upon and celebrate their learning. This part of the lesson promotes the use of mathematical language, enables the teacher to gain valuable insights into students' mathematical thinking, and reinforces the main ideas and concepts taught during the lesson.

Home Tasks

These tasks are designed to be undertaken at home. They allow students to share what they have learned in class and to reinforce their mathematical understandings with parents and/or carers. Home Tasks are provided for students in Year 3 and above.

Then give them some counters and have students create equations, investigating each of the sections. Have students record the written equation in each section. At the end, have them write either 'odd' or 'even' in each section of their charts, depending on the answers.

ODDS AND EVENS WITH LARGE NUMBERS

- For students who require a challenge, give them four different-coloured dice and an A4 sheet of paper. Have students identify a place value for each of their dice, e.g. red is thousands, blue is hundreds, green is tens and black is ones. Have students roll the dice and create a 4-digit number, then record the number on their sheet of paper. They then roll the dice again and record the second number. Have students predict whether the answer would be odd or even if the numbers were added. Have them then predict whether the answer would be odd or even if the difference was found. After the predictions have been made, have students calculate both answers and check their predictions. Note: students may need to use calculators to find/check answers.

REFLECTION

Select from the following to suit your class and their learning outcomes:

- Play the same game as in the Tuning In activity, but this time have students find the difference between the two rolled numbers on the dice. Practise finding the difference before moving on to looking at which answers are odd and which are even. Review the table created in Tuning In and see if the 'facts' students developed are holding true.
- Invite students to share their charts from the Teaching Group 'Addition and Subtraction with Counters'. Ask, 'Can you give me an example of an odd number plus an even number? Was the answer odd or even?'
- Using NTO 4.2 'Dice', have students roll two dice and just say the words 'odd' or 'even' for the answer when adding or finding the difference. This could be conducted as a class competition.

LESSON PLAN 3

TUNING IN

BUZZ

Have students stand in a circle. Select a multiple such as 5. Each student counts on by 1, and then when they reach a multiple of 5, such as 5 or 10, they say 'Buzz!' If a student fails to say 'Buzz!' at the correct time, they sit down. Continue playing a number of rounds with different multiples.

WHOLE-CLASS INTRODUCTION

MULTIPLYING ODDS AND EVENS

You will need: BLM 3 'Tables Chart 1', BLM 4 'Tables Chart 2', glue, A4 sheets of coloured paper

Enlarge BLM 3 'Tables Chart 1' and BLM 4 'Tables Chart 2', and then cut them up into individual sets of tables. Give small groups of students a set each, e.g. the 5s. Have students paste their set onto the centre of an A4 sheet of coloured paper. Ask each group to record observations on the paper about what they find when combinations of odd and even numbers from their set of tables are multiplied.

INDEPENDENT TASKS

Note: Choose from Tasks 1, 2 or 3.

You will need: BLM 5 'Odds and Evens Recording Chart', dice (one for each student), LO: *L589 'Musical number patterns: music maker'*, Student Book p. 6 'Multiplying Odd and Even Numbers'

TASK 1: MULTIPLYING DICE

Provide pairs of students with two dice. Have them roll the dice and record the numbers on BLM 5 'Odds and Evens Recording Chart'. Have them multiply the numbers and record the answer on the table, and then identify whether the answer is odd or even. Have students complete the final column of the table as a generalisation, e.g. odd × odd = odd.

TASK 2: INTERACTIVE TASK

Have students work independently on computers to explore LO: *L589 'Musical number patterns: music maker'*, making music through the exploration of number patterns.

TASK 3: STUDENT BOOK p. 6 ***'Multiplying Odd and Even Numbers'***

TEACHING GROUP

You will need: packs of playing cards

MULTIPLYING WITH PLAYING CARDS

- For students who require support, use packs of playing cards with the aces and picture cards removed. Give students a set of cards. Have them create multiplication tables, by placing cards on the desk, e.g. 5

Nelson Maths Australian Curriculum NSW 22 Teacher's Resource Book Year **4**

and 2, so 5 × 2 = 10. Have students create their own tables, and record them on a sheet of paper. Under each equation, have students identify the numbers as odd or even, and create an equation, e.g. odd × even = even. Repeat a number of times.

ODDS AND EVENS WITH DIVISION

- For students who require a challenge, have them investigate what happens with odd and even numbers and division. Pose this as an open-ended task, and ask students to create a single page with their findings, including examples.

REFLECTION

Select from the following to suit your class and their learning outcomes:

- Have students present their findings from the Whole-Class Introduction. Put similar sets of tables, e.g. the 2s, 4s and 8s, together and see if their findings were the same. Ask, 'Can we create common observations?' Create a chart, as was done in Lesson Plan 2 with addition and subtraction.
- Ask students in the Teaching Group who investigated odd and even numbers with division to present their ideas. Challenge students by asking them to 'prove' their thinking with examples.
- Replay the game of Buzz, but instead of saying 'Buzz', have students say 'Odd' or 'Even'.

Home Tasks

Select from the possible Home Tasks:

- Have students look for things around the home that have numbers, e.g. letterbox, food containers, games. Have students record the numbers and where they found them, and then circle the even numbers red and shade the odd numbers blue.
- Have students look at the newspaper (hard copy or online) for examples of odd and even numbers, e.g. results of sporting events, numbers in news stories, prices in advertisements. Have students cut out or print/save the examples and bring them to class to share.

Assessment

- Have students complete **Student Assessment p. 7**.
- Review with students **Assessment Task Card 4.1**.

During the three lessons:

- Collect created items such as the *Excel* activity in Lesson Plan 2, Independent Tasks, Task 2, and BLM 5 'Odds and Evens Recording Chart' as work samples for student portfolios.
- Review Student Book pages and note areas of difficulty.

Recommendations for Future Learning

Specific to Student Assessment p. 7; if the student is experiencing some difficulty:

Q 1 Revisit the 'Dividing Counters' activity from Lesson Plan 1 to reinforce the concepts of odd and even.
Q 2–3 A calculator could be used to find answers so the student only needs to deal with examining the answer.
Q 4 & 6 Provide copies of BLM 3 'Tables Chart 1' and BLM 4 'Tables Chart 2', so the student can practise tables and look for patterns.
Q 5 & 7 Revisit the charts and generalisations developed throughout the lessons.

If the student has not achieved the recommended skills for this unit:

1. See **Assessment Task Card 4.1** for specific recommendations.
2. Have the student work with single-digit numbers before moving to larger numbers.
3. Consolidate ideas with addition of odd and even numbers, before extending to subtraction and multiplication.
4. Review *Nelson Maths: Australian Curriculum NSW Year 3* Unit 1.

If the student has achieved the recommended skills and these skills are firmly established, consider:

1. Having the student apply the concepts to mental strategies such as open number lines, doubles and near doubles.
2. Moving forward to *Nelson Maths: Australian Curriculum NSW Year 5* Unit 3.
3. Extending the student in any of the listed activities by using larger numbers.

Unit **1** Odd and Even Numbers 23

Teaching Group

The teacher works with a small number of 'like-needs' students to teach level-appropriate concepts. The aim of this session is for the teacher to assist students at their point of need and then move them on. This is also a time to make anecdotal notes about individual students. After 10 to 20 minutes, students work on related independent tasks, while the teacher 'roves' amongst the class. The first task provided is for those students requiring further assistance, and the second task is for students who need more challenging experiences.

Assessment

Specific assessment advice is provided for teachers at the conclusion of each unit. Students should complete the Student Assessment page in the Student Book. The purpose of this page is to monitor students' understandings on the specific topic taught in the unit.

An Assessment Task Card (ATC) for each unit is also provided. This card most often provides a hands-on, investigative type of assessment task, which is less 'test focused' than the Student Assessment page. It can be used individually, with small groups or in a whole-class situation.

Recommendations for Future Learning

Each unit lists specific recommendations for future learning for both the Student Assessment page in the Student Book and the Assessment Task Card. Note: Test A provides a six-month test-based assessment covering all of the content strands taught in the first half-year period. Test B assesses content covering the whole year. Tests A and B provide teachers with a comprehensive assessment of their students' achievement levels and mathematical understandings.

STUDENT BOOK

The Student Book features engaging tasks that students can complete independently.

The Student Book includes:

Three pages of activities for each unit (one Student Book page per Lesson Plan)

The linking NSW *Mathematics K–10 Syllabus* substrand and outcome for that unit

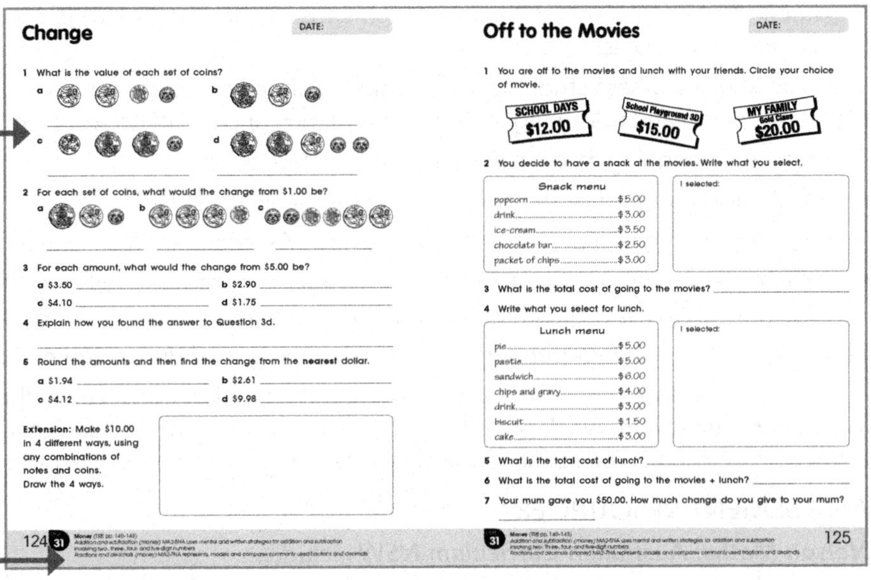

Change DATE:

1 What is the value of each set of coins?

2 For each set of coins, what would the change from $1.00 be?

3 For each amount, what would the change from $5.00 be?
a $3.50 b $2.90 c $4.10 d $1.75

4 Explain how you found the answer to Question 3d.

5 Round the amounts and then find the change from the **nearest** dollar.
a $1.94 b $2.61 c $4.12 d $9.98

Extension: Make $10.00 in 4 different ways, using any combinations of notes and coins. Draw the 4 ways.

124

Off to the Movies DATE:

1 You are off to the movies and lunch with your friends. Circle your choice of movie.
SCHOOL DAYS $12.00 School Playground 3D $15.00 MY FAMILY Gold Class $20.00

2 You decide to have a snack at the movies. Write what you select.

Snack menu	
popcorn	$5.00
drink	$3.00
ice-cream	$3.50
chocolate bar	$2.50
packet of chips	$3.00

I selected:

3 What is the total cost of going to the movies?

4 Write what you select for lunch.

Lunch menu	
pie	$5.00
pastie	$5.00
sandwich	$6.00
chips and gravy	$4.00
drink	$3.00
biscuit	$1.50
cake	$3.00

I selected:

5 What is the total cost of lunch?

6 What is the total cost of going to the movies + lunch?

7 Your mum gave you $50.00. How much change do you give to your mum?

125

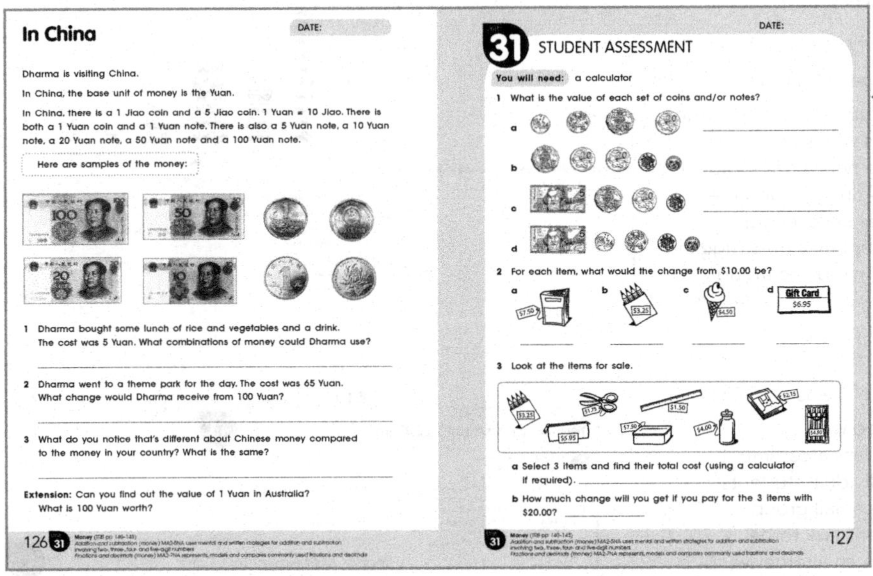

In China DATE:

Dharma is visiting China.

In China, the base unit of money is the Yuan.

In China, there is a 1 Jiao coin and a 5 Jiao coin. 1 Yuan = 10 Jiao. There is both a 1 Yuan coin and a 1 Yuan note. There is also a 5 Yuan note, a 10 Yuan note, a 20 Yuan note, a 50 Yuan note and a 100 Yuan note.

Here are samples of the money:

1 Dharma bought some lunch of rice and vegetables and a drink. The cost was 5 Yuan. What combinations of money could Dharma use?

2 Dharma went to a theme park for the day. The cost was 65 Yuan. What change would Dharma receive from 100 Yuan?

3 What do you notice that's different about Chinese money compared to the money in your country? What is the same?

Extension: Can you find out the value of 1 Yuan in Australia? What is 100 Yuan worth?

126

31 STUDENT ASSESSMENT DATE:

You will need: a calculator

1 What is the value of each set of coins and/or notes?

2 For each item, what would the change from $10.00 be?

3 Look at the items for sale.

a Select 3 items and find their total cost (using a calculator if required).

b How much change will you get if you pay for the 3 items with $20.00?

127

One Student Assessment page per unit. This page enables both student and teacher to monitor and assess an individual student's mathematical understandings throughout the units of work. Students can complete the page independently or with teacher assistance. The resulting data will assist teachers with future planning and can be used as part of a student's portfolio (for teacher–parent interviews).

Also included at the back of the Student Book is a Maths Glossary.

NELSON TEACHING OBJECTS AND LEARNING OBJECTS

Nelson Maths: Australian Curriculum integrates Information and Communication Technology (ICT) into the program. It provides up to 59 Nelson Teaching Objects (NTOs) for use on interactive whiteboards, individual computers and interactive LCD screens. The NTOs:

- illustrate mathematical concepts explicitly
- engage students actively in their own learning
- scaffold student learning
- require students to use their mathematical understandings in an engaging and meaningful context.

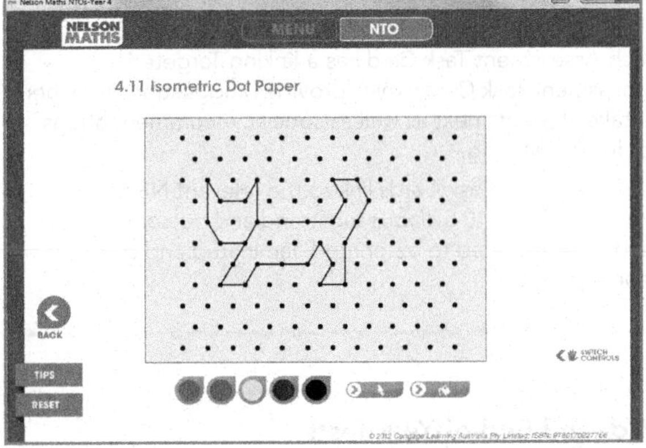

Also referred to are the pedagogically sound interactive Learning Objects (LOs) created by Education Services Australia (ESA). These are incorporated into Independent Tasks in the unit Lesson Plans. Teachers can access the LOs at:

http://www.scootle.edu.au/ec/p/home

Assessment and Planning

Assessment is a vital and ongoing part of the teaching and learning experience. Teachers need to continually monitor and assess students so their learning needs can be met and targeted when planning future educational experiences.

Teachers need to assess students' mathematical understandings (verbal and recorded), fluency, problem-solving and reasoning abilities, confidence and ability to work with others.

Assessment is ongoing, but it can be specifically targeted:

- at the middle or end of the school year (Test A and Test B)
- at the beginning of a unit (Tuning In, previous Assessment Task Card)
- at the end of a unit (Student Assessment page, Assessment Task Card).

Assessment can occur daily during mathematics sessions as part of the Whole-Class Introduction, Teaching Groups and Independent Tasks, while roving, and during Reflection.

It is not only the assessment that is important, but also what occurs after the assessment. ***Nelson Maths: Australian Curriculum NSW*** provides comprehensive and specific suggestions for students requiring further assistance as well as for those who require more challenging experiences.

ASSESSMENT RESOURCES

Nelson Maths: Australian Curriculum NSW includes the following assessment resources:

Student Assessment page (one page per unit)

This page links to Recommendations for Future Learning in the Teacher's Resource Book to guide teachers with strategic or targeted responses to the specific learning needs of individual students. The Student Assessment page is finely targeted – the questions on this page assess particular skills, enabling teachers to explicitly identify where a student may be struggling. Teachers can then respond to that specific need accordingly.

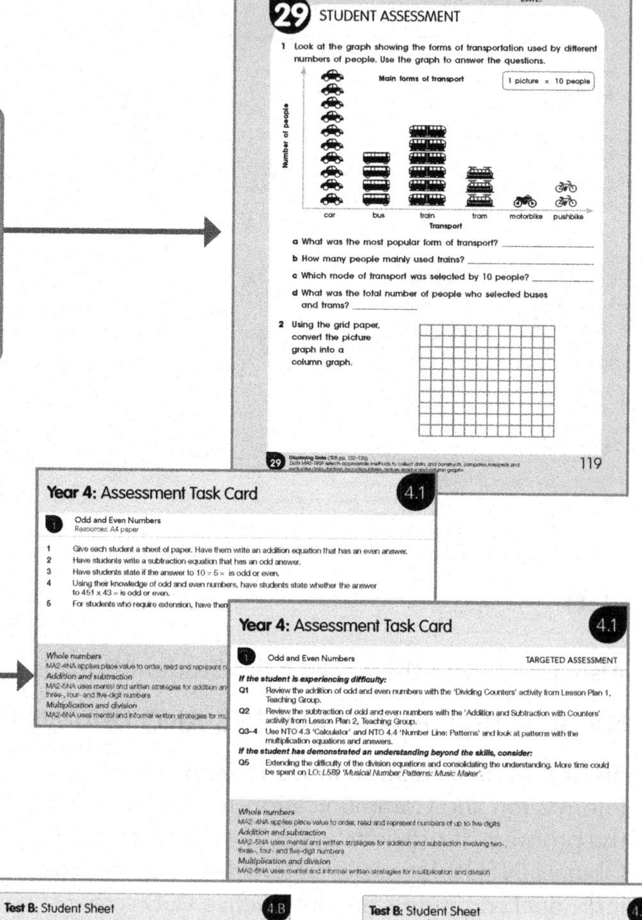

Assessment Task Cards (one for each of the 30 to 33 units)

These cards have been designed to be used with individuals, in small groups or as a whole-class assessment task. Teachers may choose to withdraw individual students to assess their understandings, work with a small group or instruct the whole class on the assessment task to be completed. The Assessment Task Cards are an alternative to pen-and-paper tests and provide some students with a more accurate way to demonstrate what they understand.

The Assessment Task Cards can be used before commencing a unit to assess students' prior knowledge, but are mostly intended to be used after the completion of a unit to assess students' understandings.

Each Assessment Task Card has a linking Targeted Assessment Task Card, which provides suggestions for where to take students next as well as specific recommendations for future learning.

All Assessment Task Cards link to the relevant NSW *Mathematics K–10 Syllabus* substrand and outcome. The cards are designed to be printed, laminated and stored for repeated use.

Mid- and End-of-Year Tests (Tests A and B)

These two paper-based tests assess students' understandings across the content strands. They are designed for mid- and end-of-year assessment. However, teachers may choose to use, for example, Year 4 Test B at the beginning of Year 5.

Test B: Student Sheet 4.B

Angles

Two-dimensional space

Chance

Test B: Student Sheet 4.B

Data

The Classroom Environment

What teachers know about learning:

- Students learn in different ways.
- Students learn from and with others.
- Students should have an opportunity to record in different ways.
- Students can follow different pathways to reach the same learning milestones.
- Students need to be encouraged to take risks in their learning.
- Students learn by being actively involved.
- Students learn by teaching others.
- Students need time to grasp the concepts introduced; that is, time for thinking, reflecting, reacting and sharing the processes and answers in mathematics.
- Students learn when they feel good about themselves – healthy self-esteem is critical.

It is important that the classroom environment:

- is stimulating and engaging and enables students to become fully immersed in learning.
- incorporates a variety of rich resources that engage students in making mathematical connections.
- is rich in the language of mathematics.
- provides students with access to calculators.
- provides students with access to Information Technology, which is an integral part of our world.
- has rich displays where students can make metacognitive links to their mathematics learning.
- is positive and encouraging and enables students to take risks in their learning.

These ideas will help to create a stimulating mathematics classroom:

- Provide step-by-step charts so students can work independently to resolve problems.
- Display students' work. Allow time to reflect upon and view completed mathematical tasks.
- Ensure students are aware of the mathematical material available to them.
- Display individual students' expertise so others can learn from them.
- Encourage students to work cooperatively in small groups.
- Integrate mathematics across the strands.
- Use charts to display helpful information and promote mathematical language.
- Include students in decisions about upcoming mathematical learning.
- Encourage peer tutoring.
- Integrate ICT into the everyday learning environment.
- Illustrate mathematics in real-world contexts.

Year 4 Overview

Unit	AC Strand/NSW Mathematics K–10 Syllabus Substrand & Outcome/AC Content Description & Code	NTOs	BLMs
Unit 1 Odd and Even Numbers	**Number and Algebra** *Whole numbers* MA2-4NA applies place value to order, read and represent numbers of up to five digits *Addition and subtraction* MA2-5NA uses mental and written strategies for addition and subtraction involving two-, three-, four- and five-digit numbers *Multiplication and division* MA2-6NA uses mental and informal written strategies for multiplication and division Investigate and use the properties of odd and even numbers (ACMNA071) AC	4.1 Hundred Chart 4.2 Dice 4.3 Calculator 4.4 Number Line: Patterns	BLM 1 Hundred Chart BLM 2 Odd and Even Equations BLM 3 Tables Chart 1 BLM 4 Tables Chart 2 BLM 5 Odds and Evens Recording Chart
Unit 2 Numbers to Tens of Thousands	**Number and Algebra** *Whole numbers* MA2-4NA applies place value to order, read and represent numbers of up to five digits Recognise, represent and order numbers to at least tens of thousands (ACMNA072) AC	4.4 Number Line: Patterns	BLM 1 Hundred Chart BLM 6 Numbers and Words 1 BLM 7 Numbers and Words 2 BLM 8 Numbers and Words 3 BLM 9 5-Digit Place-Value Chart
Unit 3 Place Value	**Number and Algebra** *Whole numbers* MA2-4NA applies place value to order, read and represent numbers of up to five digits Apply place value to partition, rearrange and regroup numbers to at least tens of thousands to assist calculations and solve problems (ACMNA073) AC	4.5 Place-Value Mat Millions 4.6 MAB: Thousands Mat 4.7 Number Expander	BLM 6 Numbers and Words 1 BLM 7 Numbers and Words 2 BLM 8 Numbers and Words 3 BLM 9 5-Digit Place-Value Chart BLM 10 Think Board BLM 11 5-Digit Number Expander

Unit	AC Strand/NSW Mathematics K–10 Syllabus Substrand & Outcome/AC Content Description & Code	NTOs	BLMs
Unit 4 Length and Temperature	**Measurement and Geometry** *Length* MA2-9MG measures, records, compares and estimates lengths, distances and perimeters in metres, centimetres and millimetres, and measures, compares and records temperatures Use scaled instruments to measure and compare lengths, masses, capacities and temperatures (ACMMG084) AC	4.8 Thermometer	BLM 12 Map of Australia BLM 13 Collecting Weather Data
Unit 5 Mass and Capacity	**Measurement and Geometry** *Volume and capacity* MA2-11MG measures, records, compares and estimates volumes and capacities using litres, millilitres and cubic centimetres *Mass* MA2-12MG measures, records, compares and estimates the masses of objects using kilograms and grams Use scaled instruments to measure and compare lengths, masses, capacities and temperatures (ACMMG084) AC		
Unit 6 Number Sequences: 3s, 6s and 9s	**Number and Algebra** *Multiplication and division* MA2-6NA uses mental and informal written strategies for multiplication and division Investigate number sequences involving multiples of 3, 4, 6, 7, 8, and 9 (ACMNA074) AC	4.1 Hundred Chart 4.4 Number Line: Patterns	BLM 1 Hundred Chart BLM 10 Think Board BLM 14 Y Chart BLM 15 Calculating Football Scores
Unit 7 Number Sequences: 4s, 8s and 7s	**Number and Algebra** *Multiplication and division* MA2-6NA uses mental and informal written strategies for multiplication and division Investigate number sequences involving multiples of 3, 4, 6, 7, 8, and 9 (ACMNA074) AC	4.1 Hundred Chart 4.4 Number Line: Patterns	BLM 1 Hundred Chart BLM 16 Number Cards: 4s and 8s BLM 33 Circles Template
Unit 8 Regular Shapes	**Measurement and Geometry** *Two-dimensional space* MA2-15MG manipulates, identifies and sketches two-dimensional shapes, including special quadrilaterals, and describes their features Compare the areas of regular and irregular shapes by informal means (ACMMG087) AC		BLM 17 1 cm Grid Paper BLM 18 Regular and Irregular Shapes

Unit	AC Strand/NSW Mathematics K–10 Syllabus Substrand & Outcome/AC Content Description & Code	NTOs	BLMs
Unit 9 2D Shapes and 3D Objects	**Measurement and Geometry** *Three-dimensional space* MA2-14MG makes, compares, sketches and names three-dimensional objects, including prisms, pyramids, cylinders, cones and spheres, and describes their features *Two-dimensional space* MA2-15MG manipulates, identifies and sketches two-dimensional shapes, including special quadrilaterals, and describes their features Make models of three-dimensional objects and describe key features (Extension of ACMMG063) AC Compare and describe two-dimensional shapes that result from combining and splitting common shapes, with and without the use of digital technologies (ACMMG088) AC	4.9 2D Shapes and 3D Objects 4.11 Isometric Dot Paper 4.19 Multilink Blocks	BLM 17 1 cm Grid Paper BLM 19 Tangram Instructions BLM 20 Tangram Shapes to Make BLM 21 Tangram Template BLM 22 Making a Tangram with a Grid BLM 23 More Tangrams! BLM 32 Isometric Dot Paper
Unit 10 Multiplication Facts (Times Tables)	**Number and Algebra** *Multiplication and division* MA2-6NA uses mental and informal written strategies for multiplication and division Recall multiplication facts up to 10 × 10 and related division facts (ACMNA075) AC	4.1 Hundred Chart 4.4 Number Line: Patterns	BLM 3 Tables Chart 1 BLM 4 Tables Chart 2 BLM 17 1 cm Grid Paper BLM 24 Rolling Tables
Unit 11 Multiplication Facts and Related Division Facts	**Number and Algebra** *Multiplication and division* MA2-6NA uses mental and informal written strategies for multiplication and division Recall multiplication facts up to 10 × 10 and related division facts (ACMNA075) AC	4.1 Hundred Chart	BLM 3 Tables Chart 1 BLM 4 Tables Chart 2 BLM 25 Matching Equations
Unit 12 Mapping	**Measurement and Geometry** *Position* MA2-17MG uses simple maps and grids to represent position and follow routes, including using compass directions Use simple scales, legends and directions to interpret information contained in basic maps (ACMMG090) AC		BLM 17 1 cm Grid Paper
Unit 13 Addition and Subtraction	**Number and Algebra** *Addition and subtraction* MA2-5NA uses mental and written strategies for addition and subtraction involving two-, three-, four- and five-digit numbers Apply place value to partition, rearrange and regroup numbers to at least tens of thousands to assist calculations and solve problems (ACMNA073) AC	4.4 Number Line: Patterns	BLM 6 Numbers and Words 1 BLM 7 Numbers and Words 2 BLM 8 Numbers and Words 3

Unit	AC Strand/NSW Mathematics K–10 Syllabus Substrand & Outcome/AC Content Description & Code	NTOs	BLMs
Unit 14 Perimeter and Area	**Measurement and Geometry** *Length* MA2-9MG measures, records, compares and estimates lengths, distances and perimeters in metres, centimetres and millimetres, and measures, compares and records temperatures *Area* MA2-10MG measures, records, compares and estimates areas using square centimetres and square metres Compare objects using familiar metric units of area and volume (ACMMG290) AC		BLM 17 1 cm Grid Paper BLM 26 Square Dot Paper BLM 27 Island Challenge
Unit 15 Multiplication and Division Strategies	**Number and Algebra** *Multiplication and division* MA2-6NA uses mental and informal written strategies for multiplication and division Develop efficient mental and written strategies and use appropriate digital technologies for multiplication and for division where there is no remainder (ACMNA076) AC	4.10 Arrays	BLM 3 Tables Chart 1 BLM 4 Tables Chart 2 BLM 25 Matching Equations BLM 28 Arrays 1 BLM 29 Arrays 2
Unit 16 More Multiplication and Division Strategies	**Number and Algebra** *Multiplication and division* MA2-6NA uses mental and informal written strategies for multiplication and division Develop efficient mental and written strategies and use appropriate digital technologies for multiplication and for division where there is no remainder (ACMNA076) AC	4.5 Place-Value Mat: Millions	BLM 3 Tables Chart 1 BLM 4 Tables Chart 2 BLM 30 Multiplication Grid
Unit 17 Volume and Capacity	**Measurement and Geometry** *Volume and capacity* MA2-11MG measures, records, compares and estimates volumes and capacities using litres, millilitres and cubic centimetres Compare objects using familiar metric units of area and volume (ACMMG290) AC	4:19 Multilink Blocks	
Unit 18 Decimals to Two Decimal Places	**Number and Algebra** *Fractions and decimals* MA2-7NA represents, models and compares commonly used fractions and decimals Recognise that the place-value system can be extended to tenths and hundredths. Make connections between fractions and decimal notation (ACMNA079) AC	4.5 Place-Value Mat: Millions	BLM 34 Decimal Numbers and Words 1 BLM 35 Decimal Numbers and Words 2 BLM 43 Symbols

Unit	AC Strand/NSW Mathematics K–10 Syllabus Substrand & Outcome/AC Content Description & Code	NTOs	BLMs
Unit 19 Chance	**Statistics and Probability** *Chance* MA2-19SP describes and compares chance events in social and experimental contexts Describe possible everyday events and order their chances of occurring (ACMSP092) AC Identify everyday events where one cannot happen if the other happens (ACMSP093) AC Identify events where the chance of one will not be affected by the occurrence of the other (ACMSP094) AC	4.2 Dice 4.12 Spinners 4.13 Coin Flip	BLM 36 Chance Words BLM 37 Circles Template
Unit 20 Collecting Data	**Statistics and Probability** *Data* MA2-18SP selects appropriate methods to collect data, and constructs, compares, interprets and evaluates data displays, including tables, picture graphs and column graphs Select and trial methods for data collection, including survey questions and recording sheets (ACMSP095) AC		BLM 38 Venn Diagrams
Unit 21 Number Patterns	**Number and Algebra** *Patterns and algebra* MA2-8NA generalises properties of odd and even numbers, generates number patterns, and completes simple number sentences by calculating missing values Explore and describe number patterns resulting from performing multiplication (ACMNA081) AC	4.1 Hundred Chart 4.4 Number Line: Patterns	BLM 1 Hundred Chart BLM 30 Multiplication Grid
Unit 22 Angles	**Measurement and Geometry** *Angles* MA2-16MG identifies, describes, compares and classifies angles Compare angles and classify them as equal to, greater than or less than a right angle (ACMMG089) AC	4.15 Angles: Measuring	BLM 59 Angles in Triangles BLM 60 Drawing Angles BLM 61 Timely Angles
Unit 23 Equivalent Fractions	**Number and Algebra** *Fractions and decimals* MA2-7NA represents, models and compares commonly used fractions and decimals Investigate equivalent fractions used in contexts (ACMNA077) AC	4.16 Comparing Shapes	BLM 30 Multiplication Grid BLM 39 Fraction Cards BLM 40 Fraction Wall BLM 41 Build Your Own Fraction Wall

Unit	AC Strand/NSW Mathematics K–10 Syllabus Substrand & Outcome/AC Content Description & Code	NTOs	BLMs
Unit 24 Counting with Fractions	**Number and Algebra** *Fractions and decimals* MA2-7NA represents, models and compares commonly used fractions and decimals Count by quarters, halves and thirds, including with mixed numerals. Locate and represent these fractions on a number line (ACMNA078) AC		BLM3 Tables Chart 1 BLM 4 Tables Chart 2 BLM 39 Fraction Cards BLM 42 Fraction Cards: Mixed Numbers and Improper Fractions BLM 43 Symbols
Unit 25 Time	**Measurement and Geometry** *Time* MA2-13MG reads and records time in one-minute intervals and converts between hours, minutes and seconds Convert between units of time (ACMMG085) AC	4.3 Calculator 4.17 Clocks: Advanced	BLM 44 Clocks BLM 45 Months
Unit 26 Time Problems	**Measurement and Geometry** *Time* MA2-13MG reads and records time in one-minute intervals and converts between hours, minutes and seconds Use am and pm notation and solve simple time problems (ACMMG086) AC		BLM 12 Map of Australia BLM 46 am and pm Times BLM 47 24-Hour Times BLM 48 Train Timetable BLM 49 Plane Flights
Unit 27 Fractions and Decimals	**Number and Algebra** *Fractions and decimals* MA2-7NA represents, models and compares commonly used fractions and decimals Recognise that the place-value system can be extended to tenths and hundredths. Make connections between fractions and decimal notation (ACMNA079) AC	4.14 Hundreds Chart: FDP	BLM 34 Decimal Numbers and Words 1 BLM 35 Decimal Numbers and Words 2 BLM 50 Hundreds Squares
Unit 28 Number Sentences	**Number and Algebra** *Patterns and algebra* MA2-8NA generalises properties of odd and even numbers, generates number patterns, and completes simple number sentences by calculating missing values Use equivalent number sentences involving addition and subtraction to find unknown quantities (ACMNA083) AC	4.14 Hundreds Chart: FDP	BLM 51 Number Cards

Unit	AC Strand/NSW Mathematics K–10 Syllabus Substrand & Outcome/AC Content Description & Code	NTOs	BLMs
Unit 29 Displaying Data	**Statistics and Probability** *Data* MA2-18SP selects appropriate methods to collect data, and constructs, compares, interprets and evaluates data displays, including tables, picture graphs and column graphs Construct suitable data displays, with and without the use of digital technologies, from given or collected data. Include tables, column graphs and picture graphs where one picture can represent many data values (ACMSP096) AC		BLM 17 1 cm Grid Paper
Unit 30 Interpreting Data	**Statistics and Probability** *Data* MA2-18SP selects appropriate methods to collect data, and constructs, compares, interprets and evaluates data displays, including tables, picture graphs and column graphs Evaluate the effectiveness of different displays in illustrating data features including variability (ACMSP097) AC		BLM 17 1 cm Grid Paper BLM 52 Insect Column Graph BLM 53 Fruit Graph
Unit 31 Money	**Number and Algebra** *Addition and subtraction (money)* MA2-5NA uses mental and written strategies for addition and subtraction involving two-, three-, four- and five-digit numbers *Fractions and decimals (money)* MA2-7NA represents, models and compares commonly used fractions and decimals Solve problems involving purchases and the calculation of change to the nearest five cents with and without digital technologies (ACMNA080) AC	4.3 Calculator 4.18 Money from Other Countries	BLM 54 Coins BLM 55 Banknotes
Unit 32 Word Problems	**Number and Algebra** *Patterns and algebra* MA2-8NA generalises properties of odd and even numbers, generates number patterns, and completes simple number sentences by calculating missing values Solve word problems by using number sentences involving multiplication or division where there is no remainder (ACMNA082) AC	4.10 Arrays	BLM 17 1 cm Grid Paper
Unit 33 Patterns	**Measurement and Geometry** *Two-dimensional space* MA2-15MG manipulates, identifies and sketches two-dimensional shapes, including special quadrilaterals, and describes their features Create symmetrical patterns, pictures and shapes with and without digital technologies (ACMMG091) AC	4.9 2D and 3D Shapes	BLM 17 1 cm Grid Paper BLM 56 5 mm Grid Paper BLM 57 Tessellation Pattern 1 BLM 58 Tessellation Pattern 2 BLM 62 Symmetrical Pictures

Nelson Maths: Australian Curriculum NSW
Scope and Sequence across the Year Levels

Note: the Working Mathematically outcomes of *Communicating*, *Problem Solving* and *Reasoning* are integrated throughout the activities and tasks in the program.

NSW Mathematics K–10 Syllabus strand and substrand	Kindergarten	Year 1	Year 2	Year 3	Year 4	Year 5	Year 6
Number and Algebra *Whole numbers*	Unit 1 Numbers to 5 Unit 2 Counting to 5 Unit 3 Groups of Things Unit 5 More Counting Unit 6 Dot Patterns Unit 8 Numbers to 10 Unit 9 Counting with Numbers to 10 Unit 10 Ten Frames Unit 11 Counting and Comparing Groups Unit 13 Ordinal Number Unit 16 Understanding More About Numbers to 10 Unit 22 Numbers Beyond 10 Unit 24 More About Numbers to 20	Unit 1 Recognising Numbers to 20 Unit 2 Counting to 20 Unit 5 Modelling Numbers Unit 6 Number Lines Unit 8 Numbers Beyond 20 Unit 15 2-digit Numbers Unit 16 More About 2-digit Numbers Unit 20 Money Unit 21 More About Money Unit 28 Place Value	Unit 1 Counting Unit 2 Modelling Numbers Unit 4 Numbers up to 1000 Unit 17 Money	Unit 1 Numbers! Numbers! Numbers! Unit 2 Numbers to 10 000 Unit 3 More About Numbers to 10 000 Unit 8 Place Value Unit 9 More About Place Value Unit 32 Money	Unit 1 Odd and Even Numbers Unit 2 Numbers to Tens of Thousands Unit 3 Place Value Unit 31 Money	Unit 1 Place Value and BODMAS Unit 3 Factors and Multiples Unit 6 Estimation	Unit 1 Place Value and BODMAS Unit 3 Integers Unit 6 Prime Numbers Unit 7 Composite Numbers Unit 18 Problems with Positive and Negative Numbers
Addition and subtraction	Unit 17 Beginning Addition Unit 19 More About Addition Unit 27 Subtraction Unit 28 More About Subtraction Unit 30 Addition, Subtraction and Money	Unit 12 Developing Mental Strategies for Addition Unit 25 Subtraction Unit 26 More About Subtraction Unit 30 Addition and Subtraction	Unit 4 Numbers Up Unit 5 Strategies for Addition Unit 6 More Strategies for Addition Unit 9 Solving Problems with Addition Unit 11 Strategies for Subtraction Unit 12 Subtraction Unit 15 More About Subtraction Unit 16 Addition and Subtraction	Unit 5 Mental Strategies for Addition Unit 6 Addition Unit 8 Place Value Unit 12 Mental Strategies for Subtraction Unit 13 Subtraction Unit 14 Connections Between Addition and Subtraction Unit 15 Solving Addition and Subtraction Problems	Unit 13 Addition and Subtraction Unit 31 Money	Unit 2 Addition and Subtraction Unit 6 Estimation Unit 31 Financial Plans	Unit 2 All Four Operations

NSW Mathematics K–10 Syllabus strand and substrand	Kindergarten	Year 1	Year 2	Year 3	Year 4	Year 5	Year 6
Number and Algebra *Multiplication and division*	Unit 20 Grouping and Sharing	Unit 3 Skip Counting Unit 14 Grouping and Sharing	Unit 23 Multiplication Unit 24 More About Multiplication Unit 27 Division Unit 28 More About Division	Unit 18 Mental Strategies for Multiplication Unit 19 More About Mental Strategies for Multiplication Unit 22 Multiplication Unit 24 Division Unit 25 More About Division	Unit 6 Number Sequences: 3s, 6s and 9s Unit 7 Number Sequences 4s, 8s and 7s Unit 10 Multiplication Facts (Times Tables) Unit 11 Multiplication Facts and Related Division Facts Unit 15 Multiplication and Division Strategies Unit 16 More Multiplication and Division Strategies	Unit 3 Factors and Multiples Unit 7 Multiplication of Large Numbers A Unit 10 Multiplication of Large Numbers B Unit 11 Division with Remainders Unit 15 Mental Strategies	Unit 2 All Four Operations
Fractions and decimals	Unit 23 Halves	Unit 13 Halves and Quarters	Unit 7 Fractions Unit 30 More About Fractions	Unit 26 Fractions Unit 27 More About Fractions Unit 28 Decimals	Unit 18 Decimals to 2 Decimal Places Unit 23 Equivalent Fractions Unit 24 Counting with Fractions Unit 27 Fractions and Decimals	Unit 13 Addition and Subtraction of Decimals Unit 17 Fractions Unit 18 Decimals to Three Places Unit 23 Addition of Fractions Unit 24 Subtraction of Fractions Unit 27 Fractions and Decimals	Unit 12 Decimal Representations of the Metric System Unit 14 Addition and Subtraction of Decimals Unit 15 Multiplication of Decimals Unit 17 Fractions Unit 23 Addition of Fractions Unit 24 Subtraction of Fractions Unit 27 Fractions of a Quantity Unit 28 Fractions, Decimals and Percentages Unit 31 Percentage Discounts
Patterns and algebra	Unit 14 Patterns	Unit 18 Patterns Unit 19 Number Patterns	Unit 18 Number Patterns Unit 19 More Number Patterns	Unit 21 Patterns	Unit 21 Number Patterns Unit 28 Number Sentences Unit 32 Word Problems	Unit 21 Number Patterns Unit 28 Number Sentences	Unit 11 Cartesian System Unit 21 Number Sequences

NSW Mathematics K–10 Syllabus strand and substrand	Kindergarten	Year 1	Year 2	Year 3	Year 4	Year 5	Year 6
Measurement and Geometry *Length* *Area* *Volume and capacity* *Mass*	Unit 7 Length and Area Unit 26 How Much Does It Hold? Unit 21 Mass	Unit 10 Length and Area Unit 29 Capacity Unit 23 Mass	Unit 3 Length Unit 25 Area Unit 20 Capacity Unit 13 Mass	Unit 4 Length Unit 23 Area Unit 30 Capacity Unit 10 Mass	Unit 4 Length and Temperature Unit 14 Perimeter and Area Unit 17 Volume Unit 5 Mass and Capacity	Unit 4 Length, Area and Volume Unit 14 Perimeter Unit 16 Area Unit 5 Mass and Capacity	Unit 4 Area and Perimeter Unit 13 Length and Area Problems Unit 16 Volume and Capacity Unit 5 Mass and Capacity
Time	Unit 15 Time Unit 29 More About Time	Unit 9 Time Unit 22 More AboutTime	Unit 14 Telling the Time Unit 21 More About Time	Unit 16 Time	Unit 25 Time Unit 26 Time Problems	Unit 25 Time Unit 26 Time Problems	Unit 25 Time Unit 26 Timetables and Timelines
Three-dimensional space *Two-dimensional space*	Unit 18 More About Shapes and Objects Unit 12 2D Shapes	Unit 17 3D Objects Unit 4 2D Shapes	Unit 26 3D Objects Unit 8 Transformation with 2D Shapes	Unit 31 3D Objects Unit 29 Symmetry	Unit 9 Drawings of 3D Objects Unit 8 Regular Shapes Unit 33 Patterns	Unit 9 3D Objects Unit 8 Shapes Unit 32 Transformations	Unit 8 2D Shapes and 3D Objects Unit 9 Prisms and Pyramids Unit 8 2 Shapes and 3D Objects Unit 32 Transformations Unit 33 Use of Transformations
Position	Unit 4 Position	Unit 7 Position	Unit 10 Position	Unit 7 Position	Unit 12 Mapping	Unit 12 Grid References	Unit 10 Mapping/Grid References
Angles				Unit 17 Angles	Unit 22 Angles	Unit 22 Angles Unit 33 Angle Applications	Unit 22 Angles
Statistics and Probability *Chance*		Unit 24 Chance	Unit 29 Chance	Unit 20 Chance	Unit 19 Chance	Unit 19 Chance	Unit 19 Chance
Data	Unit 25 Our Class	Unit 27 Data	Unit 22 Data	Unit 11 Our Community – Data	Unit 20 Collecting Data Unit 29 Displaying Data Unit 30 Interpreting Data	Unit 20 Collecting Data Unit 29 Data Displays Unit 30 Interpreting Data	Unit 20 Collecting Data Unit 29 Data Displays Unit 30 Interpreting Data

Odd and Even Numbers

Number and Algebra

Whole numbers MA2-4NA applies place value to order, read and represent numbers of up to five digits

Addition and subtraction MA2-5NA uses mental and written strategies for addition and subtraction involving two-, three-, four- and five-digit numbers

Multiplication and division MA2-6NA uses mental and informal written strategies for multiplication and division

addition, counting sequence, division, even, integers, multiplication, numbers, odd, pattern, subtraction, Venn diagram

LESSON PLAN 1

TUNING IN

COUNTING PATTERNS

You will need: NTO 4.1 'Hundred Chart' or an enlarged copy of BLM 1 'Hundred Chart'

Display NTO 4.1 'Hundred Chart' on the IWB. Give a student a starting number and a counting sequence. Have the student, with the help of the class, complete the counting sequence, either by selecting squares or pointing to numbers on the chart. Have the class count the sequence aloud. Ask questions, such as, 'What is the starting number of the counting sequence? What are we counting by? What patterns can we see with the numbers?' Explore counting sequences, such as by 2s, 3s and 5s, with different starting points.

WHOLE-CLASS INTRODUCTION

EXPLORING ODD AND EVEN NUMBERS

You will need: an IWB blank screen or poster paper, whiteboard markers

On the IWB or poster paper, draw a Venn diagram with the headings 'Odd Numbers' and 'Even Numbers' for each circle. As a class, have students brainstorm and record what they know about odd and even numbers in the relevant circles and the aspects that are common. This may include descriptions, examples of numbers, where these numbers may be found, and why odd and even numbers are important.

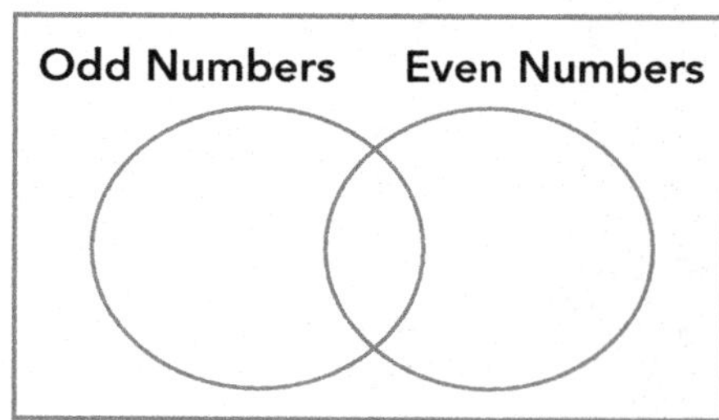

INDEPENDENT TASKS

Note: Choose from Tasks 1, 2 or 3.

You will need: long rolls of paper, LO: *L1064 'Musical number patterns: odds and evens'*, Student Book p. 4: 'Identifying Odd and Even Numbers'

TASK 1: CONTINUE THE SEQUENCE

Provide each student with a long roll of paper. Give students an even starting number, differentiated according to student needs. Have students write the starting number on the roll of paper and then continue writing the next 10 to 20 even numbers. Repeat this activity for odd numbers. Then have students form pairs to compare their sequences of even and odd numbers, looking for similarities and differences.

TASK 2: INTERACTIVE TASK

Have students work independently on computers on LO: *L1064 'Musical number patterns: odds and evens'*, completing a counting rule that matches a pattern on a number line, based on odd or even numbers.

TASK 3: STUDENT BOOK p. 4 *'Identifying Odd and Even Numbers'*

TEACHING GROUP

You will need: large sheets of paper with two separate large circles drawn on them, counters, accessto the internet

DIVIDING COUNTERS

- For students who require support, provide large sheets of paper with two separate large circles drawn on them. Give each student a handful of counters, and ask them to divide the counters evenly into the two circles. Explore with students that if there is an even number of counters, no counters will be left over, but if there is an odd number of counters, one counter will be left over. Give students more counters and ask them to count and record the total number of counters, then predict if the number is odd or even. Then have them place the counters in the circles and clarify their answer. Repeat a number of times.

ODD AND EVEN NUMBER EXPLORATION

- For students who require a challenge, provide them with a number set, e.g. decimals or negative numbers, and have them use the internet to find out if these numbers are odd or even. Have students collect information to share with the rest of the class – this may be in the form of a *Word* document or links that could be shared on the IWB.

REFLECTION

Select from the following to suit your class and their learning outcomes:

- Return to the Venn diagram created in the Whole-Class Introduction, and review and add new ideas about odd and even numbers. Display the chart in the classroom.
- Have students share the number sequences they made in 'Continue the Sequence', Independent Tasks, Task 1. Invite students to share what they found when they made the comparison with their partner.
- Invite students in the Teaching Group who completed the internet research to share what they found. This could be added to the Venn diagram of odd and even numbers.

LESSON PLAN 

TUNING IN

TWO DICE

You will need: two large dice or NTO 4.2 'Dice', a large sheet of paper

Have students sit in a circle and take it in turns to roll the two dice and add the numbers together. Once students understand the activity, ask them to note whether the numbers rolled are odd or even and if they add to an odd or even number. The information could be recorded in a table on the board or on a large sheet of paper. At the end of the activity, have students draw conclusions, e.g. odd + odd = even.

WHOLE-CLASS INTRODUCTION

ADDING ODDS AND EVENS

You will need: cards with 2-digit numbers (these can be created from BLM 1 'Hundred Chart' or from pieces of card), calculators or NTO 4.3 'Calculator'

Give each student a different 2-digit number – distribute both odd and even numbers. Ask students to use their numbers to create an even number, with addition or subtraction, by finding a partner. Have them check their answer, either with pen and paper or a calculator (or NTO 4.3 'Calculator'). Equations could be recorded on the board as a group. Then have students pair up to create odd numbers, again recording equations on the board as a group. Challenge students by having them form groups of three to create odd and even numbers. Link this activity to the chart that students developed in the Tuning In activity. Ask, 'Do the rules we came up with work for 2-digit numbers?' Add any conclusions regarding subtraction to the chart, e.g. odd – even = odd.

INDEPENDENT TASKS

Note: Choose from Tasks 1, 2 or 3.

You will need: BLM 2 'Odd and Even Equations', *Excel*, Student Book p. 5 'Adding and Subtracting Odd and Even Numbers'

TASK 1: ODD AND EVEN EQUATIONS

Have students complete BLM 2 'Odd and Even Equations'.

TASK 2: INTERACTIVE TASK

Have students create an even counting pattern of five numbers down a column in *Excel*. They can then extend the pattern by selecting their five numbers and dragging the bottom right corner of the selected box down the page to continue the number pattern. Have students explore different patterns. Challenge students to explore patterns such as odd, even, odd, even.

TASK 3: STUDENT BOOK p. 5 *'Adding and Subtracting Odd and Even Numbers'*

TEACHING GROUP

You will need: A4 sheets of paper, rulers, counters, four different-coloured dice, calculators

ADDITION AND SUBTRACTION WITH COUNTERS

- For students who require support, give them a sheet of A4 paper. Have them divide the paper into quarters and label as shown right.

odd and odd	odd and even
even and odd	even and even

Then give them some counters and have students create equations, investigating each of the sections. Have students record the written equation in each section. At the end, have them write either 'odd' or 'even' in each section of their charts, depending on the answers.

ODDS AND EVENS WITH LARGE NUMBERS

- For students who require a challenge, give them four different-coloured dice and an A4 sheet of paper. Have students identify a place value for each of their dice, e.g. red is thousands, blue is hundreds, green is tens and black is ones. Have students roll the dice and create a 4-digit number, then record the number on their sheet of paper. They then roll the dice again and record the second number. Have students predict whether the answer would be odd or even if the numbers were added. Have them then predict whether the answer would be odd or even if the difference was found. After the predictions have been made, have students calculate both answers and check their predictions. Note: students may need to use calculators to find/check answers.

REFLECTION

Select from the following to suit your class and their learning outcomes:

- Play the same game as in the Tuning In activity, but this time have students find the difference between the two rolled numbers on the dice. Practise finding the difference before moving on to looking at which answers are odd and which are even. Review the table created in Tuning In and see if the 'facts' students developed are holding true.
- Invite students to share their charts from the Teaching Group 'Addition and Subtraction with Counters'. Ask, 'Can you give me an example of an odd number plus an even number? Was the answer odd or even?'
- Using NTO 4.2 'Dice', have students roll two dice and just say the words 'odd' or 'even' for the answer when adding or finding the difference. This could be conducted as a class competition.

LESSON PLAN 3

TUNING IN

BUZZ

Have students stand in a circle. Select a multiple such as 5. Each student counts on by 1, and then when they reach a multiple of 5, such as 5 or 10, they say 'Buzz!' If a student fails to say 'Buzz!' at the correct time, they sit down. Continue playing a number of rounds with different multiples.

WHOLE-CLASS INTRODUCTION

MULTIPLYING ODDS AND EVENS

You will need: BLM 3 'Tables Chart 1', BLM 4 'Tables Chart 2', glue, A4 sheets of coloured paper

Enlarge BLM 3 'Tables Chart 1' and BLM 4 'Tables Chart 2', and then cut them up into individual sets of tables. Give small groups of students a set each, e.g. the 5s. Have students paste their set onto the centre of an A4 sheet of coloured paper. Ask each group to record observations on the paper about what they find when combinations of odd and even numbers from their set of tables are multiplied.

INDEPENDENT TASKS

Note: Choose from Tasks 1, 2 or 3.

You will need: BLM 5 'Odds and Evens Recording Chart', dice (one for each student), LO: *L589 'Musical number patterns: music maker'*, Student Book p. 6 'Multiplying Odd and Even Numbers'

TASK 1: MULTIPLYING DICE

Provide pairs of students with two dice. Have them roll the dice and record the numbers on BLM 5 'Odds and Evens Recording Chart'. Have them multiply the numbers and record the answer on the table, and then identify whether the answer is odd or even. Have students complete the final column of the table as a generalisation, e.g. odd × odd = odd.

TASK 2: INTERACTIVE TASK

Have students work independently on computers to explore LO: *L589 'Musical number patterns: music maker'*, making music through the exploration of number patterns.

TASK 3: STUDENT BOOK p. 6 ***'Multiplying Odd and Even Numbers'***

TEACHING GROUP

You will need: packs of playing cards

MULTIPLYING WITH PLAYING CARDS

- For students who require support, use packs of playing cards with the aces and picture cards removed. Give students a set of cards. Have them create multiplication tables, by placing cards on the desk, e.g. 5

and 2, so 5 × 2 = 10. Have students create their own tables, and record them on a sheet of paper. Under each equation, have students identify the numbers as odd or even, and create an equation, e.g. odd × even = even. Repeat a number of times.

ODDS AND EVENS WITH DIVISION

- For students who require a challenge, have them investigate what happens with odd and even numbers and division. Pose this as an open-ended task, and ask students to create a single page with their findings, including examples.

REFLECTION

Select from the following to suit your class and their learning outcomes:

- Have students present their findings from the Whole-Class Introduction. Put similar sets of tables, e.g. the 2s, 4s and 8s, together and see if their findings were the same. Ask, 'Can we create common observations?' Create a chart, as was done in Lesson Plan 2 with addition and subtraction.
- Ask students in the Teaching Group who investigated odd and even numbers with division to present their ideas. Challenge students by asking them to 'prove' their thinking with examples.
- Replay the game of Buzz, but instead of saying 'Buzz', have students say 'Odd' or 'Even'.

Home Tasks

Select from the possible Home Tasks:

- Have students look for things around the home that have numbers, e.g. letterbox, food containers, games. Have students record the numbers and where they found them, and then circle the even numbers red and shade the odd numbers blue.
- Have students look at the newspaper (hard copy or online) for examples of odd and even numbers, e.g. results of sporting events, numbers in news stories, prices in advertisements. Have students cut out or print/save the examples and bring them to class to share.

Assessment

- Have students complete **Student Assessment p. 7**.
- Review with students **Assessment Task Card 4.1**.

During the three lessons:

- Collect created items such as the *Excel* activity in Lesson Plan 2, Independent Tasks, Task 2, and BLM 5 'Odds and Evens Recording Chart' as work samples for student portfolios.
- Review Student Book pages and note areas of difficulty.

Recommendations for Future Learning

Specific to Student Assessment p. 7; if the student is experiencing some difficulty:

Q 1 Revisit the 'Dividing Counters' activity from Lesson Plan 1 to reinforce the concepts of odd and even.

Q 2–3 A calculator could be used to find answers so the student only needs to deal with examining the answer.

Q 4 & 6 Provide copies of BLM 3 'Tables Chart 1' and BLM 4 'Tables Chart 2', so the student can practise tables and look for patterns.

Q 5 & 7 Revisit the charts and generalisations developed throughout the lessons.

If the student has not achieved the recommended skills for this unit:

1. See **Assessment Task Card 4.1** for specific recommendations.
2. Have the student work with single-digit numbers before moving to larger numbers.
3. Consolidate ideas with addition of odd and even numbers, before extending to subtraction and multiplication.
4. Review *Nelson Maths: Australian Curriculum NSW Year 3* Unit 1.

If the student has achieved the recommended skills and these skills are firmly established, consider:

1. Having the student apply the concepts to mental strategies such as open number lines, doubles and near doubles.
2. Moving forward to *Nelson Maths: Australian Curriculum NSW Year 5* Unit 3.
3. Extending the student in any of the listed activities by using larger numbers.

Numbers to Tens of Thousands

Number and Algebra

Whole numbers MA2-4NA applies place value to order, read and represent numbers of up to five digits

digits, hundreds, hundreds of thousands, millions, number lines, numbers, place value, tens, tens of thousands, thousands, units

LESSON PLAN 

TUNING IN

TRAFFIC LIGHTS

You will need: a whiteboard, three different-coloured whiteboard markers (ideally red, orange and green), sticky notes

Thousands	Hundreds	Tens	Ones

Draw the table at right on the whiteboard. On a sticky note, write a 4-digit number using four different digits, e.g. 4567 or 3218 but not 4422. Students take it in turns to guess the number. If they guess the correct digit in the correct place-value column, they receive a green circle, which is written on the whiteboard in green marker pen. If they guess a digit that is in the number, but in the wrong place-value column, they receive an orange circle. If they guess a digit that does not appear in the number, they receive a red circle. Make a note of the number. Students then guess again, trying to narrow down the answer. Each guess is recorded down the board, leading to the four green lights. Students could play in teams or against you. The activity could be extended by including tens or hundreds of thousands.

WHOLE-CLASS INTRODUCTION

NUMBERS AND WORDS

You will need: BLM 6 'Numbers and Words 1', BLM 7 'Numbers and Words 2', BLM 8 'Numbers and Words 3', modelling equipment (e.g. MAB, Unifix blocks, counters, craft sticks)

Enlarge and photocopy BLM 6 'Numbers and Words 1', BLM 7 'Numbers and Words 2' and BLM 8 'Numbers and Words 3', and cut up into individual components, enough for one for each student. Give each student a component, and have them silently find their partner. Once they have found their partner, have them make their number with equipment they select. Have groups show and share their number – in words and made with the equipment.

INDEPENDENT TASKS

Note: Choose from Tasks 1, 2 or 3.

You will need: BLM 6 'Numbers and Words 1', BLM 7 'Numbers and Words 2', BLM 8 'Numbers and Words 3', LO: *L867 'Wishball: whole numbers'*, Student Book p. 8 'Matching Words and Numerals'

TASK 1: MEMORY OR SNAP

Give each student a copy of BLM 6 'Numbers and Words 1', BLM 7 'Numbers and Words 2' and BLM 8 'Numbers and Words 3' to create their own game of 'Memory' or 'Snap'. Have students decorate and cut out the cards, and then work in small groups to play their preferred game. It may be possible to copy the BLM on coloured paper to minimise decorating time.

TASK 2: INTERACTIVE TASK

Have students work independently on computers to play LO: *L867 'Wishball: whole numbers'*, where students add or subtract numbers according to place value as they work towards a target number. This can also be played as a whole-class activity.

TASK 3: STUDENT BOOK p. 8 *'Matching Words and Numerals'*

TEACHING GROUP

You will need: BLM 6 'Numbers and Words 1', BLM 7 'Numbers and Words 2', BLM 8 'Numbers and Words 3', BLM 9 '5-Digit Place-Value Chart', LO: *L871 'Wishball challenge: whole numbers'*

MISSING WORDS

- For students who require support, use BLM 6 'Numbers and Words 1', BLM 7 'Numbers and Words 2' and BLM 8 'Numbers and Words 3' with the words blanked out, and BLM 9 '5-Digit Place-Value Chart'. Write the first number on BLM 6 (4023) in the correct columns of the place-value chart. Show students how this can help them to identify the components of the number, then help them to write the numbers in words. It may be necessary to scaffold students with a word list on the board. Have students work through the number list, writing the numbers on the place-value chart and attempting to write the numbers in words.

ODD AND EVEN NUMBER EXPLORATION

- For students who require a challenge, use LO: *L871 'Wishball challenge: whole numbers'*, which allows students to extend LO: *L867 'Wishball: whole numbers'*.

REFLECTION

Select from the following to suit your class and their learning outcomes:

- Play the traffic light activity from Tuning In, extending the numbers to the tens of thousands.
- Have students comment on the game they played in 'Memory or Snap', Independent Tasks, Task 1. Ask, 'Did you find the activity challenging? What were the tricky numbers?'
- Play LO: *L867 'Wishball: whole numbers'* as a class, perhaps with students competing against you.

TUNING IN

I AM THINKING OF A NUMBER

You will need: sticky notes

Write a 5-digit number on a sticky note (e.g. 32 564). Give students a range, e.g. 'I am thinking of a number between 32 000 and 33 000.' Students take turns guessing the number, with only 'higher' or 'lower' being provided as the answer to each guess. Students need to state the number correctly, e.g. 'thirty-two thousand, three hundred and fifty', otherwise the turn is counted as a miss. The aim is to guess the number in the least amount of guesses. Students could play against you, or in teams. Play a number of times.

WHOLE-CLASS INTRODUCTION

ORDERING 5-DIGIT NUMBERS

You will need: cards with 5-digit numbers (these can be created from BLM 6 'Numbers and Words 1', BLM 7 'Numbers and Words 2' and BLM 8 'Number Words 3')

Give each student a different card. You may wish to use only the number cards initially, and then vary the activity by including the word cards. Have students arrange themselves in order from smallest to largest according to their cards. Have students read out the cards to determine if the order is correct. Discuss the order and how it was determined.

INDEPENDENT TASKS

Note: Choose from Tasks 1, 2 or 3.

You will need: glue, poster paper, BLM 6 'Numbers and Words 1', BLM 7 'Numbers and Words 2', BLM 8 'Numbers and Words 3', *PowerPoint*, Student Book p. 9 'Ordering Numbers'

TASK 1: ORDERING NUMBERS POSTERS

Provide small groups of students with glue, poster paper and a selection of number and word cards, according to students' abilities. These could be created from BLM 6 'Numbers and Words 1', BLM 7 'Numbers and Words 2' and BLM 8 'Numbers and Words 3'. Have students order the cards and create a poster to display.

TASK 2: INTERACTIVE TASK

Provide students with a selection of 5-digit numbers and ask them to create a *PowerPoint* presentation, placing the numbers in order. Have them include the relevant number in words, and an image to represent the number. This activity could be completed individually or in pairs.

TASK 3: STUDENT BOOK, p. 9 *'Ordering Numbers'*

TEACHING GROUP

You will need: LO: *L2317 'Number trains'*, sticky notes, glue, large sheets of paper

NUMBER TRAINS

- For students who require support, have them work together at the IWB (or independently on computers), using LO: *L2317 'Number trains'* to order the numbers on the train carriages from smallest to largest to complete the train.

EXTENDING THE RANGE

- For students who require a challenge, give them a wider range of numbers to order – these could include 6- or 7-digit numbers, or decimals. Provide students with number cards only – these could be created on sticky notes to meet individual needs. Have students order the sticky notes and stick them to a large sheet of paper. Then have students write the numbers in words under each of the notes. Ask students to check each other's work.

REFLECTION

Select from the following to suit your class and their learning outcomes:

- Play the game 'I Am Thinking of a Number' from the Tuning In activity, but give a wider range to enable students to practise their elimination skills.
- Invite groups to share the posters they made in 'Ordering Numbers Posters', Independent Tasks, Task 1. Ask questions, such as, 'How well did you work together as a group? Which numbers did you find tricky to place? Why did you decide to present your work like this?'
- Do the ordering activity from the Whole-Class Introduction again, but instead of having students hold the cards, attach the numbers to their backs so they cannot see their own number, and then have them complete the ordering in silence, by working together as a group.
- Have students share their *PowerPoint* presentations from Independent Tasks, Task 1.

LESSON PLAN

TUNING IN

CHALK NUMBER LINES

You will need: chalk, a large working space

Draw a blank number line on the ground in chalk. Note: it may be possible to use the school's basketball or netball court lines. Give one student a number, e.g. 11 256. Ask them where they would stand on the number line and why. Give another student a different number, e.g. 21 325. Ask them where they would stand and allow them to adjust the scaling on the number line. Continue with a number of students, examining both the order of the numbers and the scaling of the number line. Note: numbers could be prepared in advance to cater for students' abilities.

WHOLE-CLASS INTRODUCTION

ORDERING 5-DIGIT NUMBERS TO SCALE

You will need: cards with 5-digit numbers (these can be created from BLM 6 'Numbers and Words 1', BLM 7 'Numbers and Words 2' and BLM 8 'Numbers and Words 3') a large working space, sticky notes, chalk, a digital or flip camera

Repeat the activity 'Ordering 5-Digit Numbers' from Lesson Plan 2, but when the numbers have been ordered, have students place the numbers on a scaled number line. Students will need a large space to work in (possibly outside or in the corridor). Students could use sticky notes or chalk to help create their number line. When complete, either photograph (with a digital camera) or video (with a flip camera) the number line.

INDEPENDENT TASKS

Note: Choose from Task 1, 2 or 3

You will need: BLM 6 'Numbers and Words 1', BLM 7 'Numbers and Words 2', BLM 8 'Number Words 3, long rolls of paper, glue, NTO 4.4 'Number Line: Patterns', Student Book p. 10 'Number Lines and 5-Digit Numbers'

TASK 1: LONG NUMBER LINES

Give small groups of students a set of number cards in words only (these could be created from BLM 6 'Numbers and Words 1', BLM 7 'Numbers and Words 2' and BLM 8 'Number Words 3') and a long roll of paper. Have students create a scaled number line with their numbers. They may wish to rewrite their number words into digits before starting, but their presented number line needs to be in words.

TASK 2: INTERACTIVE TASK

Have students explore and use NTO 4.4 'Number Line: Patterns', either individually on computers or in small groups on the IWB. A specific activity could be provided or the task could be open-ended. Students could be give a starting number, a scale and a group of numbers to place accurately on the number line.

TASK 3: STUDENT BOOK p. 10 *'Number Lines and 5-Digit Numbers'*

TEACHING GROUP

You will need: NTO 4.4 'Number Line: Patterns', long rolls of paper, BLM 1 'Hundred Chart', calculators

ELECTRONIC NUMBER LINE

- For students who require support, use NTO 4.4 'Number Line: Patterns' with small groups of students working on the IWB. Provide students with starting numbers and ranges, and have them order the numbers correctly on the number line. Use sets of smaller numbers, i.e. less than 100 and then less than 1000. Use materials such as Hundred Charts (BLM 1 'Hundred Chart' could be used) to scaffold students.

WIDENING THE SCALE

- For students who require a challenge, provide a wider selection of numbers, i.e. millions, and have them create scaled number lines. Students may need a calculator to help work out the value of suitable divisions.

REFLECTION

Select from the following to suit your class and their learning outcomes:

- Share the photos or video taken during the Whole-Class Introduction. Have students comment on the numbers and patterns that they observe in the footage.
- Have students present any of the scaled number lines created in the Independent Tasks or the Teaching Group. Ask questions, such as, 'What was difficult about this activity? How did you work out the divisions?'
- Display the number lines around the classroom. Have students make comparisons between the number lines, and say what they liked about them.

Home Tasks

Select from the possible Home Tasks:

- Ask students to look for numbers written in words around the home, e.g. in books, in instruction manuals, on packaging, in newspapers. Have them create a list and bring it to school. Collect the lists in a common place, such as on the IWB or a poster.
- Ask students to find five 5-digit numbers at home. Have them record the numbers and where they were found, and write the number in words.

Assessment

- Have students complete **Student Assessment p. 11**.
- Review with students **Assessment Task Card 4.2**.

During the three lessons:

- Collect the digital materials, such as the *PowerPoint* presentation made in Lesson Plan 2, Independent Tasks, Task 2, or the photos/video taken in Lesson Plan 3, Whole-Class Introduction, and add them to students' digital portfolios.
- Make a note of students who have completed the Home Tasks. Comment on areas of difficulty students may have experienced.
- Review Student Book pages and note areas of difficulty.

Recommendations for Future Learning

Specific to Student Assessment p. 11; if the student is experiencing some difficulty:

Q 1–3 Revisit the activities from Lesson Plan 1 where students match words and numbers.

Q 4 Revisit the ordering activities from Lesson Plan 2. This could be scaffolded with the use of the place-value chart (BLM 9 '5-Digit Place-Value Chart').

Q 5–6 Revisit the process of ordering numbers from smallest to largest, then create a number line. Note: numbers could be written on sticky notes, or created from BLM 6 'Numbers and Words 1', BLM 7 'Numbers and Words 2' and BLM 8 'Numbers and Words 3'.

If the student has not achieved the recommended skills for this unit:

1. See **Assessment Task Card 4.2** for specific recommendations.
2. Have the student work with single-digit numbers before moving to larger numbers.
3. Consolidate ideas about number lines with smaller numbers, e.g. 2-digit numbers, to help with the understanding of scale.
4. Review *Nelson Maths: Australian Curriculum NSW Year 3* Units 2 and 3.

If the student has achieved the recommended skills and these skills are firmly established, consider:

1. Having the student create a game, pairing numbers and words.
2. Moving forward to *Nelson Maths: Australian Curriculum NSW Year 5* Unit 1.
3. Extending the student in any of the listed activities by using larger numbers, moving into millions.

Unit 3 Place Value

Number and Algebra

Whole numbers MA2-4NA applies place value to order, read and represent numbers of up to five digits

expanded notation, hundreds, number expander, numbers, patterns, place value, rename, tens, tens of thousands, thousands, units

LESSON PLAN 1

TUNING IN

PLACE-VALUE CHART

You will need: NTO 4.5 'Place-Value Mat: Millions', NTO 4.6 'MAB: Thousands Mat'

Draw a place-value chart on the board or use NTO 4.5 'Place-Value Mat: Millions' on the IWB. Discuss the columns and what happens between them. Provide examples or a list of numbers for students to enter into the chart. Note: NTO 4.6 'MAB: Thousands Mat' has a visual option to help this discussion.

WHOLE-CLASS INTRODUCTION

INTO THE PARTS

You will need: NTO 4.5 'Place-Value Mat: Millions'

On the board, explore with students how to write numbers in their place-value components. For example: 245 = 200 + 40 + 5. Begin with smaller numbers and work up to numbers with at least five digits. Link this to the place-value chart from the Tuning In activity (either on the board or with NTO 4.5 'Place-Value Mat: Millions'). Then reverse the process, providing students with 30000 + 2000 + 700 + 1 and have them provide the number – 32701. Explore a variety of numbers, depending on students' abilities.

INDEPENDENT TASKS

Note: Choose from Tasks 1, 2 or 3.

You will need: BLM 6 'Numbers and Words 1', BLM 7 'Numbers and Words 2', BLM 8 'Numbers and Words 3', BLM 10 'Think Board', modelling equipment (e.g. MAB, Unifix blocks, counters), a digital camera, *Word*, *Excel*, Student Book p. 12 'Hidden Numbers'

TASK 1: THINK BOARD

Give students a 5-digit number (BLM 6 'Numbers and Words 1', BLM 7 'Numbers and Words 2' and BLM 8 'Numbers and Words 3' could be used). Give each student a copy of BLM 10 'Think Board'. Have students write the number in the centre of the think board. Then have students work individually to complete the four sections of the think board with the following:

1. Write the number in words.
2. Write the expanded place-value format (e.g. 200 + 30 + 1).
3. Write a word problem that includes the number.
4. Make the number with equipment, e.g. MAB. Take a photo of the modelled equipment, print it and add it to the think board, or have the student draw what they modelled.

TASK 2: INTERACTIVE TASK

Have students work independently on computers, using a program such as *Word* or *Excel*, to create their own place-value chart. They need to create a version where they can enter a variety of numbers. When complete, have students form pairs and give each other numbers to enter into the place-value charts.

TASK 3: STUDENT BOOK p. 12 *'Hidden Numbers'*

TEACHING GROUP

You will need: NTO 4.6 'MAB: Thousands Mat', MAB

PLACE-VALUE CHART WITH MAB

- For students who require support, use NTO 4.6 'MAB: Thousands Mat' with MAB. Have students explore the break-up of numbers into the expanded format. Work with small numbers, e.g. 325. Draw out the links between the equipment and the numeric representation. Invite students to the IWB to show representations.

LARGER NUMBERS IN EXPANDED FORM

- For students who require a challenge, provide larger numbers into the hundreds of thousands, or millions, and have them write the expanded form for each number. Have them check each other's work. Students could be further extended by looking at decimal numbers.

REFLECTION

Select from the following to suit your class and their learning outcomes:

- Have students share their place-value charts from Independent Tasks, Task 2, and explain how they developed the charts.
- Have students share and explain their think boards from Independent Tasks, Task 1.
- Revisit NTO 4.6 'MAB: Thousands Mat'. Students could work towards the target number in the NTO activity, building on their place-value skills.

LESSON PLAN 2

TUNING IN

EXPANDED FORM

You will need: a set of 5-digit numbers (BLM 6 'Numbers and Words 1', BLM 7 'Numbers and Words 2' and BLM 8 'Numbers and Words 3' could be used), sticky notes, one sheet of paper per group

Provide groups of students with a 5-digit number (or a number to suit their ability). Have them break the number into its expanded form, with each part on a different sticky note. For example: 32 450 = 30 000 + 2000 + 400 + 50 where the 30 000 is on one sticky note, 2000 is on another and so on. Then on the sheet of paper, have groups record the different ways the expanded form could be written, using the sticky notes to help. For example:

32 450 = 30 000 + 2000 + 400 + 50

32 450 = 30 000 + 50 + 400 + 2000

Students could be creative with their presentation. Discuss the advantage of writing the equation based on place value, i.e. 30 000 + 2000 + 400 + 50. Discuss what happens if there is no value for a place. Ask, 'What do we write?'

WHOLE-CLASS INTRODUCTION

EXPLORING NUMBER EXPANDERS

You will need: a constructed number expander (use BLM 11 '5-Digit Number Expander')

Show students the number expander and write a number on it. Ask questions, such as, 'What number can we see? What is the number now? Has it changed?' Record comments and ideas on the board. Discuss the term 'rename' and what this means.

INDEPENDENT TASKS

Note: Choose from Tasks 1, 2 or 3.

You will need: BLM 11 '5-Digit Number Expander', adhesive plastic book covering, scissors, NTO 4.7 'Number Expander', Student Book p. 13 'Spider Web'

TASK 1: MY NUMBER EXPANDER

Have students make their own number expander using BLM 11 '5-Digit Number Expander'. You could cover the number expanders with adhesive plastic so they can be written on and erased (note: it is difficult to fold the number expanders if they are laminated; if they are covered with plastic, they will be easier to fold). Have students work in pairs to use their number expanders, giving each other numbers to explore.

TASK 2: INTERACTIVE TASK

Explore NTO 4.7 'Number Expander'. Students could work on computers, entering numbers given by you and observing what happens. Alternatively, students could work in pairs, giving each other numbers.

TASK 3: STUDENT BOOK, p. 13 ***'Spider Web'***

TEACHING GROUP

You will need: NTO 4.7 'Number Expander', constructed number expanders

NUMBER TRAINS

- For students who require support, have them work with their number expanders. Provide smaller numbers, expand the numbers and have students record them. This could be supported with the use of NTO 4.7 'Number Expander'. This could also be reversed, with you providing expanded forms, e.g. 300 + 80 + 2, and students finding the non-expanded form on their number expander (e.g. 382).

EXTENDING THE RANGE

- For students who require a challenge, provide a number such as 74 hundreds and 21 ones, and have them investigate all the different variations of renaming. Extend the numbers into the hundreds of thousands, and millions. Have students record the number and all its variations on a sheet of paper. Students could investigate a range of numbers.

REFLECTION

Select from the following to suit your class and their learning outcomes:

- Give students a number and have them record it on their number expander, and then have them offer the number renamed in different ways. Record this on the board.
- Use NTO 4.7 'Number Expander' to check answers provided by students.
- Invite students to share the larger numbers they investigated in the Teaching Group 'Extending the Range'.

LESSON PLAN 3

TUNING IN

WISHBALL

You will need: LO: *L867 'Wishball: whole numbers'*

As a class, play LO: *L867 'Wishball: whole numbers'* on the IWB, or have students play on computers. As students are playing, look at the setup of the page – the place-value columns and the fact that they are multiples of 10 – as well as the abacus and scaling allowing for different representations of numbers. Discuss the impact of this on the game.

WHOLE-CLASS INTRODUCTION

REGROUPING NUMBERS

You will need: NTO 4.6 'MAB: Thousands Mat'

Have students create numbers using the MAB function on NTO 4.6 'MAB: Thousands Mat'. Record this as an equation, e.g. 400 + 20 + 1. Using the arrows at the top of the place-value columns, have students break down the MAB or combine it to regroup the number. Record the variation, e.g. 300 + 120 + 1, and discuss how the number is the same but renamed. Make links to the columns being multiples of 10. Repeat this with a variety of numbers.

INDEPENDENT TASKS

Note: Choose from Tasks 1, 2 or 3.

You will need: newspapers, sheets of paper, scissors, glue, constructed number expanders (from Lesson Plan 2), LO: *L1999 'Scale matters: tens of thousands'*, Student Book p. 14 'Expanding Numbers'

TASK 1: NEWSPAPER NUMBERS

Give small groups of students newspapers and have them find 5-, 6- or 7-digit numbers, according to their abilities. Have them cut out the numbers and stick them on a sheet of paper. Under each number, have students break the number into its place-value components, e.g. 60 000 + 1000 + 500 + 60 + 3. To extend the activity, have students rename the number and the expansion. Students may like to use their number expanders from Lesson Plan 2. Set a total of numbers students need to find and stick on their paper. Have students retain the pieces of paper.

TASK 2: INTERACTIVE TASK

Have students work independently on computers, using LO: *L1999 'Scale matters: tens of thousands'* to locate numbers on a number line. Note: there are various versions of this Learning Object to cater for different abilities.

TASK 3: STUDENT BOOK p. 14 *'Expanding Numbers'*

TEACHING GROUP

You will need: newspapers, sheet of paper, glue, scissors, BLM 9 '5-Digit Place-Value Chart', LO: *L8629 'Scale matters: all numbers: assessment'*, constructed number expander (from Lesson Plan 2)

NEWSPAPER NUMBER HUNT

- For students who require support, provide them with newspapers and have them find numbers according to their ability, e.g. 3-digit numbers. Have students cut out the numbers and stick them on a sheet of paper. Then have students write the numbers on BLM 9 '5-Digit Place-Value Chart'. If possible, extend students to expand the numbers with the aid of the place-value chart and their number expander from Lesson Plan 2 and record this on their sheet of paper.

SCALE MATTERS ASSESSMENT

- For students who require a challenge, have them work independently on computers to complete LO: *L8629 'Scale matters: all numbers: assessment'*. This allows students to use the process of renaming while trying to locate numbers on a number line. This assessment provides a report for the user.

REFLECTION

Select from the following to suit your class and their learning outcomes:

- Have students share their newspaper numbers either from Independent Tasks, Task 1, or the Teaching Group. Ask students to provide the number and the expansion. Ask the rest of the group what other ways the numbers could be renamed.
- Revisit NTO 4.6 'MAB: Thousands Mat'. Ask students to identify the place-value columns and what values they hold. Look at what happens when the arrows are pressed. Ask questions, such as, 'What happens to the columns? Where do the blocks go? Does the total value of the number change?' Draw out the links between the multiples and division of 10 between columns.

Home Tasks

Select from the possible Home Tasks:

- Have students take their number expanders home to share with parents or carers. Students should explain how the number expander works, and show some examples.
- Have students find two more numbers in their local paper, a newspaper or an online paper to add to their 'Newspaper Numbers' from Lesson Plan 3, Independent Tasks, Task 1. Have students bring the numbers to class and add to their previous collection.

Assessment

- Have students complete **Student Assessment p. 15**.
- Review with students **Assessment Task Card 4.3**.

During the three lessons:

- Collect digital materials, such as the place-value chart students created in Lesson Plan 1, Independent Tasks, Task 2, and add them to students' digital portfolios.
- Collect students' think boards from Lesson Plan 1, Independent Tasks, Task 1. Make note of any areas students may have had difficulty with. Add the think boards to student portfolios.
- Collect the work from 'Newspaper Numbers' in Lesson Plan 3, Independent Tasks, Task 1, as evidence of in-class work and homework, and the ability to work in groups.

Recommendations for Future Learning

Specific to Student Assessment p. 15; if the student is experiencing some difficulty:

Q 1 & 6 Use NTO 4.5 'Place-Value Mat: Millions' to review the columns, values and representations in a place-value chart.

Q 2 & 5 Use the student's constructed number expander from Lesson Plan 2 to explore the naming and renaming of numbers. This could also be revisited using NTO 4.7 'Number Expander'.

Q 3–4 Revisit the writing of expanded form, scaffolded with NTO 4.5 'Place-Value Mat: Millions'.

If the student has not achieved the recommended skills for this unit:

1. See **Assessment Task Card 4.3** for specific recommendations.
2. Have the student work with smaller numbers before moving to larger numbers.
3. Link the use of the place-value chart with the number expander and the written expanded form. Reinforce using similar language between all three elements. Examples could be displayed in the classroom.
4. Review *Nelson Maths: Australian Curriculum NSW Year 3* Units 3 and 8.

If the student has achieved the recommended skills and these skills are firmly established, consider:

1. Having the student complete 'Renaming Numbers' from *Nelson Maths Building Mental Strategies Big Book 4*, pp. 12–13.
2. Moving forward to *Nelson Maths: Australian Curriculum NSW Year 5* Unit 1.
3. Extending the student in any of the listed activities by using larger numbers.

Unit 4 Length and Temperature

Measurement and Geometry

Length MA2-9MG measures, records, compares and estimates lengths, distances and perimeters in metres, centimetres and millimetres, and measures, compares and records temperatures

centimetres, cooler, length, measure, metres, millimetres, ruler, scale, tape measure, temperature, thermometer, trundle wheels, warmer

LESSON PLAN 1

TUNING IN

INFORMAL MEASUREMENT

You will need: a range of hands-on equipment (e.g. MAB, Unifix blocks, counters)

Have pairs of students select equipment to use in measuring the length of objects. Have them move around the classroom and measure the length of five objects, and record these on a sheet of paper. Have pairs share the length of one of their objects. Ask, 'What is the difficulty in measuring with equipment such as this?' (it is hard to be accurate with decimal/fractional amounts; you need a lot of equipment as they are small pieces; it is time-consuming to place the small pieces of equipment against the object being measured)

WHOLE-CLASS INTRODUCTION

RULERS

You will need: one ruler for each student, a metre ruler

Give each student a ruler. Spend time talking about the ruler. Ask ,'What do you see?' Look at elements such as scales, the way the numbers are shown (or represented with lines), whether the scale is reversed on the other side of the ruler, the similarities to a number line and where the scale actually starts, i.e. where the zero is. You may also want to compare these features to a metre ruler.

INDEPENDENT TASKS

Note: Choose from Tasks 1, 2 or 3.

You will need: pre-cut ribbon or string lengths, rulers, *Excel*, Student Book p. 16 'Lengths of Feet'

TASK 1: WORKING BACKWARDS

Before class, measure the length of a range of objects in the classroom with a piece of ribbon or string. Cut each piece to the length of the object and label it A, B, C and so on. On a sheet of paper, record the letter, the object measured and its length for your own records. Have at least enough lengths of ribbon or string for each pair of students to have one, and then perhaps another five. Give pairs of students one of the lengths and challenge them to find the object in the room that matches it. Once they think they have found the object, have them record the letter of the ribbon, the name of the object and the actual length. They return their length of ribbon to a common place and select another to repeat the activity. After a period of time, bring students together as a group, and go through each of the letters and see if anyone found the actual items. Point out how many items in a classroom have similar lengths.

TASK 2: INTERACTIVE TASK

Give students rulers and have them measure the length of 10 objects: a table, a book, their drink bottle, a pencil, the board, the longest object in the room, four objects of their own choice. Have students record the name of the object and the length in centimetres. Have students then collect the data in a table format in *Excel*. If they complete the activity early, they can compare the lengths with a classmate, and create a comparison column in their tables. Have students add a heading to their tables.

TASK 3: STUDENT BOOK p. 16 *'Lengths of Feet'*

TEACHING GROUP

You will need: rulers, objects to measure, circular objects, ribbon or string

PRACTISING MEASURING

- For students who require support, give them each a ruler and a number of objects to measure. Have students practise measuring and recording their results. Check that each student is using the ruler correctly and recording accurately. Talk about strategies such as counting on, using the scale and relating it to a number line. Remind students to include units.

CIRCULAR OBJECTS

- For students who require a challenge, have them spend time working out how to measure circular objects. Give them rulers and tell them they cannot use any other formal measuring equipment, but can they use informal materials such as ribbon or string.

REFLECTION

Select from the following to suit your class and their learning outcomes:

- Discuss with students the difference between formal and informal measurement. Ask when the different methods are appropriate. Ask, 'Which method of measurement do you prefer?'
- Have students share their *Excel* tables from Independent Tasks, Task 2. Compare answers from the set items measured and invite students to share the optional ones. Look for the longest.
- Have students share how they measured circular objects in the Teaching Group 'Circular Objects'. Ask them to explain the processes that they used.

LESSON PLAN 2

TUNING IN

ORDERING HEIGHT AND LENGTHS

Have students arrange themselves in height order from shortest to tallest. Record the order on the board. Then have them arrange themselves in length of hand order from shortest to longest. Record the order on the board. Ask, 'Is the order the same?' Have students measure the length of their arms from elbow to finger tips, and arrange themselves from shortest to longest. Record the order on the board. Ask, 'Has the order changed?' This activity could be extended by examining some of da Vinci's ideas about body proportions.

WHOLE-CLASS INTRODUCTION

EQUIPMENT TO USE WITH LENGTH

You will need: a range of measuring equipment (e.g. rulers, metre rulers, tape measures, trundle wheels)

Examine each piece of equipment and discuss its purpose. Highlight the units of measurement used with each item. Have students explain how to use equipment such as trundle wheels and tape measures, and correct any misconceptions. Students might like to share when they have used this equipment previously.

INDEPENDENT TASKS

Note: Choose from Tasks 1, 2 or 3.

You will need: a range of measuring equipment (e.g. rulers, metre rulers, tape measures, trundle wheels); paper; *PowerPoint*; access to the internet; Student Book p. 17 'Comparing Length'

TASK 1: MEASURING OUTSIDE

Have students go outside and in small groups use the measuring equipment to measure a space such as a playground, and map it on a sheet of paper. Allow students to select the tool they feel appropriate for the area they are measuring. Make sure students record the name of the area and label its length. It may be appropriate for students to work in different areas. This activity could be extended by having students measure the objects within the space, e.g. the play equipment.

TASK 2: INTERACTIVE TASK

Using software such as *PowerPoint*, have students collect or create images of objects, and add to their presentation. Under each of the images, students need to identify the tool they would use to measure the object and what units it would be in, e.g. the length of a car may be measured with a tape measure and the units would be metres. Have students complete a set of at least 10 images.

TASK 3: STUDENT BOOK p. 17 *'Comparing Length'*

TEACHING GROUP

You will need: magazines, paper, scissors, glue, rulers, objects to measure

MAGAZINE MEASUREMENT

- For students who require support, give them a magazine and ask them to find an item that would be measured in metres. Have students cut out and paste the image on a sheet of paper, and write the label 'metres'. Repeat this a number of times, looking at different units of measurement. Then students could discuss which piece of equipment would be used to complete the measurement.

GIVEN THE LENGTH

- For students who require a challenge, give them some objects and ask them to measure the length. Have students measure in centimetres, then convert the measurements into millimetres and then metres. Repeat this a number of times. Have students collect the information in a table. Note: it may be relevant to discuss conversion factors before the activity or allow students to determine this through completing the activity.

Object	Length in cm	Length in mm	Length in m

REFLECTION

Select from the following to suit your class and their learning outcomes:

- Ask students to write a reflection about which measuring tool they enjoyed using the most and why. Invite some students to share their reflections.
- Show students each of the measuring tools used in Lesson Plan 2. Ask, 'What units are used for this device? What are some of the things we can measure with this device?'
- Have students share their *PowerPoint* presentations from Independent Tasks, Task 2. Look for similarities, differences and creativity. These could be shared as a digital book.

LESSON PLAN 3

TUNING IN

WARMER, COOLER

Play the game 'Warmer, Cooler'. Select an item in the room, perhaps related to the lesson on temperature, and have two or three students move around the room trying to find the object, with your description of 'Warmer' as they get closer and 'Cooler' as they move away. The winner is the student who identifies the object. Repeat a number of times with different students.

WHOLE-CLASS INTRODUCTION

MEASURING TEMPERATURE

You will need: a thermometer, NTO 4.8 'Thermometer'

Show students a real thermometer. Ask, 'What is this used for? Can anyone explain how to use it?' Using NTO 4.8 'Thermometer', show students how to read the vertical scale. Try a number of different temperatures, with students reading the scale. Ask, 'Can you explain how you worked that out?' To extend the activity, different scales could be discussed, i.e. the Celsius and Fahrenheit scales.

INDEPENDENT TASKS

Note: Choose from Tasks 1, 2 or 3.

You will need: maps of the school, thermometers, internet access, BLM 12 'Map of Australia', Student Book p. 18 'Temperature'

TASK 1: TEMPERATURE AROUND THE SCHOOL

Give pairs of students a map of the school and a thermometer. Discuss safety issues, such as not running with the thermometer or what to do if they drop the thermometer. Have students move around the school, measuring the temperature at various inside and outside locations and recording them on the map. Prior to starting, discuss how long they should wait in each location for the temperature to stabilise before moving on.

TASK 2: INTERACTIVE TASK

Have students go to the Bureau of Meterology website. Allow them to explore the site and find the temperature for each of the capital cities of Australia. Students can record the temperatures on BLM 12 'Map of Australia'. They then find the forecast for your local area for the next five days and add it to the map. Have students comment on the back of the BLM, e.g. what they found interesting; the difference between minimum and maximum temperatures; where they would like to be and why, based on temperature. This activity could be extended by having students investigate and search for other sites that have weather information, e.g. national and state newspapers.

TASK 3: STUDENT BOOK p. 18 *'Temperature'*

TEACHING GROUP

You will need: thermometers, NTO 4.8 'Thermometer', paper, containers of water, containers of sand, magazines, newspapers, scissors, glue, poster paper

PRACTISING MEASURING TEMPERATURE

- For students who require support, give them a thermometer. Have them measure the temperature while holding the thermometer in their hand and record this on a sheet of paper. Check that each student is reading the scale correctly. Have one of students show the value on NTO 4.8 'Thermometer'. Repeat the activity with students placing the thermometer in a small container of water and then a container of sand.

IN THE PRESS

- For students who require a challenge, provide them with magazines and newspapers, and have them search the materials to find different references to temperature. This may include weather reports, news articles and advertisements. Have them cut out the image or detail and, as a group, create a collage on poster paper.

REFLECTION

Select from the following to suit your class and their learning outcomes:

- Have students share their maps of the school from Independent Tasks, Task 1. Look at variation in temperatures and differences within the same locations. Examine which areas are warmer and which are cooler.
- Invite students in the Teaching Group 'Practising Measuring Temperature' to share what they discovered. Ask questions, such as, 'Which object was warmest? Which was the coolest?'
- Have students in the Teaching Group 'In the Press' share and display the collages.

Home Tasks

Select from the possible Home Tasks:

- Have students ask their parents or carers what measuring equipment they use at home or at work. Have students create a list and bring it to class to share, e.g. rulers, thermometers, measuring cups.
- Have students record the temperature for a week using BLM 13 'Collecting Weather Data'. Missed data could be found in the paper or online.

Assessment

- Have students complete **Student Assessment p. 19**.
- Review with students **Assessment Task Card 4.4**.

During the three lessons:

- Collect the digital materials such as the *Excel* measurements from Lesson Plan 1, Independent Tasks, Task 2, or *PowerPoint* images from Lesson Plan 2, Independent Tasks, Task 2, and add to students' digital portfolios.
- Collect students' maps from Independent Tasks, Task 1, as evidence of spatial understanding and the ability to work in small groups.
- Keep students' weather data homework as a sample and evidence of homework.

Recommendations for Future Learning

Specific to Student Assessment p. 19; if the student is experiencing some difficulty:

Q 1 Revisit how to use a ruler. Spend time with the student and the equipment, measuring different items and recording lengths.

Q 2–3 Have the student sort images or objects into those measured in centimetres and those measured in metres. Have students measure the lengths of objects and order from shortest to longest.

Q 4 –5 Revisit NTO 4.8 'Thermometer' and review how to read a thermometer. Have the student show specific temperatures using the NTO, as well as reading values.

If the student has not achieved the recommended skills for this unit:

1. See **Assessment Task Card 4.4** for specific recommendations.
2. Have the student revisit informal measurement, and make direct comparisons between objects, identifying which is shorter and which is longer.
3. Have the student continue to work with the specific materials, e.g. rulers. Link scales with number lines to build on prior knowledge.
4. Review *Nelson Maths: Australian Curriculum NSW Year 3* Unit 4.

If the student has achieved the recommended skills and these skills are firmly established, consider:

1. Having the student move to working with the units of kilometres for length.
2. Moving forward to *Nelson Maths: Australian Curriculum NSW Year 5* Unit 4.
3. Having the student investigate different measurement and temperature scales used around the world.

Unit 5

Mass and Capacity

Measurement and Geometry

Volume and capacity MA2-11MG measures, records, compares and estimates volumes and capacities using litres, millilitres and cubic centimetres

Mass MA2-12MG measures, records, compares and estimates the masses of objects using kilograms and grams

capacity, grams, heaviest, hefting, kilograms, lightest, litres, mass, measure, millilitres, scales, volume, weigh

LESSON PLAN 1

TUNING IN

HEFTING

You will need: various items to heft (e.g. bean bags, potatoes)

Give small groups of students items to heft to determine which are the heaviest and which are the lightest. Have the groups report back to the class about how they determined which item was the heaviest/lightest.

WHOLE-CLASS INTRODUCTION

SCALES

You will need: different scales (e.g. balance scales, kitchen scales, bathroom scales)

Show each piece of equipment. Ask questions, such as, 'What is this piece of equipment used for? Can you explain how to use it?' Draw on students' prior knowledge and examine each piece of equipment and how to use it. Discuss the scale on the equipment and the difference between grams and kilograms.

INDEPENDENT TASKS

Note: Choose from Tasks 1, 2 or 3.

You will need: different scales (e.g. balance scales, kitchen scales, bathroom scales), various items to weigh, *Excel*, a few bags of potatoes, Student Book p. 20 'Kilograms or Grams'

TASK 1: WEIGHING ITEMS

Set up the scales around the room. Have students work in pairs to select 10 different items, and to estimate what each item would weigh. Have them record this data in a table, as below, and then record the actual weight. Emphasise the importance of estimation, and the importance of units in both the estimation and the actual weight. As students are working, move around the class to ensure they are completing the estimation. After students have completed their measurements, have them write a comment below the table, reflecting on the accuracy of their estimations.

Name of Object	Estimated Weight	Actual Weight

TASK 2: INTERACTIVE TASK 

Set up the scales around the room. Have students work in pairs and use *Excel* to create a table to record the weights of potatoes. Students need to weigh one potato and record the data, then a second potato and so on until they have recorded the weights of at least 10 potatoes. Allow students to develop their own table, recording system and processes. The activity could be extended so students create a graph of their data, or find the average weight of their potatoes.

TASK 3: STUDENT BOOK p. 20 *'Kilograms or Grams'*

TEACHING GROUP

You will need: different scales (e.g. balance scales, kitchen scales, bathroom scales), various items to weigh

PRACTISING WEIGHING

- For students who require support, give them a number of items to weigh. Students should begin by predicting whether they will use grams or kilograms, and what piece of equipment they will select. Have students place the item on the scale, and discuss how to read the scale, i.e. where to find the numbers and the units. Have students record the value of the weight on a sheet of paper, including the units.

FINDING THE TOTAL WEIGHT

- For students who require a challenge, provide them with a total, e.g. 2.5 kg. Have them work in pairs to find at least five items in the classroom to make the total exactly. Have students record the weights and the final list of items as they proceed.

REFLECTION

Select from the following to suit your class and their learning outcomes:

- Have students complete a written reflection on the potato activity in Independent Tasks, Task 1. Have them reflect on whether their processes were effective and efficient. Ask, 'What would you change if you did the activity again?'
- Have students from the Teaching Group 'Practising Weighing' share the weights of the items they weighed. Note similar items weighed (e.g. different books) and whether the weights were the same or not. Examine some of the heavier and some of the lighter items students weighed. Ask, 'What was tricky about estimating? Were you right all the time? Did that matter? Why is it important to estimate?'
- Have students in the Teaching Group 'Finding the Total Weight' share the objects they found to make the total weight. Have them discuss the approach they took to find the items.

LESSON PLAN 2

TUNING IN

UNIFIX IN CONTAINERS

You will need: Unifix blocks; various containers labelled A, B, C and so on

Have students work in pairs or small groups. Give each group a container and some Unifix blocks and have students stack the blocks into the container to find the maximum number of blocks that fit. Have students record the number of blocks on the container. Then rotate the containers to the next group and repeat. Repeat this a number of times and have students compare answers.

WHOLE-CLASS INTRODUCTION

FILLING CONTAINERS

You will need: various containers – two with the same capacity and others with different capacities, rice, a large container to work in (to reduce the mess)

Show students the containers. Explain that they are to find the two containers that have the same capacity. Review the term 'capacity' with students. Allow students to drive the activity and to solve the problem.

INDEPENDENT TASKS

Note: Choose from Tasks 1, 2 or 3.

You will need: rice, measuring jugs, various containers (labelled A, B, C and so on), large containers to work in (to reduce mess), LO: *L1993 'Squirt: two containers'*, Student Book p. 21 'Graphing Capacity'

TASK 1: MEASURING CONTAINERS

Have students use the rice and measuring jugs to determine the capacity of each container. Have them record the measurement and the label of the container in a table. To extend the activity, students could be encouraged to estimate the capacity of each of the containers before measuring.

TASK 2: INTERACTIVE TASK

Have students work independently on computers, using LO: *L1993 'Squirt: two containers'*. Students need to determine the number of fills of a particular container, which is then used to find the number of fills for a second container. Note: there are various versions of this Learning Object to cater for different abilities.

TASK 3: STUDENT BOOK p. 21 *'Graphing Capacity'*

TEACHING GROUP

You will need: various containers, Unifix blocks, rice, measuring jugs, large container to work in (to reduce mess), LO: *L1990 'Squirt: three containers'*

UNIFIX AND RICE

- For students who require support, give them a container and have them measure the capacity with Unifix blocks. Have them record the total. Observe and discuss the different methods used. Have students repeat with different containers. Then have students repeat the activity with rice and measuring jugs to determine the capacity of the containers. Have them discuss the difference between the two results.

LO: L1990 'SQUIRT: THREE CONTAINERS'

- For students who require a challenge, have them work independently on computers, using LO: *L1990 'Squirt: three containers'* to determine the capacity relationship between three containers. Note: there are various versions of this Learning Object to cater for different abilities.

REFLECTION

Select from the following to suit your class and their learning outcomes:

- Have students consider the two methods of measuring capacity – one using Unifix blocks and the other using rice. Ask, 'Which method would be appropriate to use and when? Which one is more accurate? Which method did you prefer? Why?'
- Repeat the activity from the Whole-Class Introduction, using a different set of containers, trying to find the two containers with the same capacity. Observe to see if students' methods have changed.
- Have students share their measurements for each of the labelled containers from Independent Tasks, Task 1. Review to see if all students had the same or different answers. Check to see if students used the correct units.

LESSON PLAN 3

TUNING IN

MASS OR CAPACITY?

You will need: empty containers, containers filled with rice or water, measuring jugs, rice, scales

Show students a container and ask, 'What will we measure – mass or capacity?' If it is full, have students measure mass with the scales. If it is empty, have them measure capacity with rice and a measuring jug. Once capacity is measured, then the mass can also be measured. Check that students are using the measuring equipment correctly.

WHOLE-CLASS INTRODUCTION

SCALES

You will need: measuring jugs; scales (for grams and for kilograms); various items to be measured in kilograms, grams and litres

Provide students with an amount, e.g. 3 kg. Have students select which piece of equipment would be used to complete the measurement, e.g. scales for kilograms, and then have them weigh an item and record the value. Repeat a number of times with different items to be measured in kilograms, grams or litres.

INDEPENDENT TASKS

Note: Choose from Tasks 1, 2 or 3.

You will need: various containers (labelled A, B, C and so on), rice, measuring jugs, scales, *Excel*, Student Book p. 22 'Mass and Capacity'

TASK 1: HOW MANY TO FILL?

Provide students with a range of containers. Have them explore how many of container *x* makes up container *y* in terms of capacity and mass. Have students record this information in a table, and then write it as an equation, e.g. $x = 5 \times y$. Have them explore all combinations of their set of containers.

TASK 2: INTERACTIVE TASK

Have pairs of students select a container. Have them measure the capacity and the mass of the container and record the data in a table using *Excel*. Students should note the label of the container in the table. Have them repeat this at least five times with different containers. When complete, have them examine if there is any relationship between the mass and capacity of the containers. This activity could be extended, with students using the formula functions in *Excel*.

TASK 3: STUDENT BOOK p. 22 *'Mass and Capacity'*

TEACHING GROUP

You will need: water, small and large measuring cups, small container, larger container, computers, LO: *L3534 'How high?'*, calculators

GETTING A FEEL FOR AMOUNTS

- For students who require support, give them a small measuring cup and have them measure 5 mL of water. Have them investigate how many of these would be needed to fill a small container (e.g. the size of a medicine bottle). Repeat with a larger scale, e.g. 100 mL for filling a 1 L container.

LO: L3534 'HOW HIGH?'

- For students who require a challenge, have them work independently on computers, using LO: *L3534 'How high?'* to determine the volume of one container based on the volume of another. This Learning Object can use the formula of volume, and students can use a calculator for support.

REFLECTION

Select from the following to suit your class and their learning outcomes:

- Show students a container, and have them estimate what the capacity/volume may be. Have students record their guesses on the board. Then measure and see who is the closest. Repeat with different containers. Have students whose estimates are the closest share their techniques.
- Invite students to share how many containers made up the different ones in Independent Tasks, Task 1. See which groups exhausted all of the possibilities.
- Have students share and display their *Excel* tables from Independent Tasks, Task 2. Ask, 'Was there a relationship between mass and capacity?' Have groups share their largest mass, greatest capacity and so on.

Home Tasks

Select from the possible Home Tasks:

- Have students explore their pantries or cupboards to find five items of different mass and volume. They draw the objects, record the amounts and bring the work to school to share.
- Have students watch the news or look at the newspaper to see where mass and capacity are mentioned, e.g. in advertisements or sports reports. Have students note these and add to a class poster of 'Mass and Capacity in the Media'.

Assessment

- Have students complete **Student Assessment p. 23**.
- Review with students **Assessment Task Card 4.5**.

During the three lessons:

- Collect the digital materials, such as the *Excel* tables from Lesson Plan 3, Independent Tasks, Task 2, and add to students' digital portfolios.
- Collect students' written reflections of their work to add to their portfolios.
- Take photos of students completing activities to add to their digital portfolios. Students could be invited to write a comment about the activity and their learning.

Recommendations for Future Learning

Specific to Student Assessment p. 23; if the student is experiencing some difficulty:

Q 1 Revisit hefting and then weighing objects to order from lightest to heaviest.

Q 2 Have the student sort objects into those measured with grams and those with kilograms. Show an object with a mass of 10 g and one with a mass of 1 kg. Have the student feel the difference.

Q 3 Revisit the term 'capacity' with the student.

Q 4 Show the student measuring jugs and revisit the reading of scales. Ensure the student is reading scales correctly and starting from the correct place.

Q 5 Have the student explore moving water or rice from one container to another. Have them examine how many of one container is needed to make up the other. Have students practise recording this, and ensure they are not reversing the expression.

If the student has not achieved the recommended skills for this unit:

1. See **Assessment Task Card 4.5** for specific recommendations.
2. Have the student revisit informal measurement and make direct comparisons between objects, identifying which is lighter or heavier or has the greatest capacity.
3. Have the student continue to work with the specific materials, i.e. scales and measuring jugs. Link scales with number lines to build on prior knowledge. Ensure the student reads from the correct point, and that they wait for scales to adjust to zero before they begin measuring.
4. Review *Nelson Maths: Australian Curriculum NSW Year 3* Units 10 and 30.

If the student has achieved the recommended skills and these skills are firmly established, consider:

1. Having the student move to working with tonnes.
2. Moving forward to *Nelson Maths: Australian Curriculum NSW Year 5* Unit 5.
3. Having the student investigate different mass and capacity scales used around the world.

Number Sequences: 3s, 6s and 9s

Number and Algebra

Multiplication and division MA2-6NA uses mental and informal written strategies for multiplication and division

multiple, number sequence, patterns, place value

LESSON PLAN 1

TUNING IN

HUNDRED CHART: BY 3S

You will need: NTO 4.1 'Hundred Chart' or a Hundred Chart poster

Display NTO 4.1 'Hundred Chart' or a Hundred Chart poster and have students count by the number sequence of 3, starting at a variety of points. Invite students to comment on any patterns they observe on the chart or in the numbers as they are counting. Ask questions, such as, 'How did you know the number 9 was next? What is the next number? How did you work that out?' The activity can be extended by having students explore the number sequence backwards.

WHOLE-CLASS INTRODUCTION

SEQUENCES ON NUMBER LINES

You will need: NTO 4.4 'Number Line: Patterns'

Display NTO 4.4 'Number Line: Patterns' and have students count by the number sequence of 3. Have students explore the patterns at different starting points, working both forwards and backwards. Invite students to the board to complete the number sequences. Ask questions, such as, 'What is the next number? How did you work that out? What strategy did you use? How is the number line similar/different to the Hundred Chart?'

INDEPENDENT TASKS

Note: Choose from Tasks 1, 2 or 3.

You will need: one long roll of paper per student, calculators, paper, Student Book p. 24 'Number Sequences of 3'

TASK 1: HOW FAR CAN THE SEQUENCE GO?

Give each student a long roll of paper. Provide students with a starting number, according to ability, e.g. 3 or 5 or 300 or 1322. Have students count forwards by 3s as far as they can in a certain time frame, writing as neatly as possible. Then have students swap their counting sequence and have the next student add to it within another time frame. This rotation could be conducted a number of times. Ask students, 'How do we know if the sequence is correct?' Have students share answers. Then have the last student of the rotation peer-assess the work. Display the rolls.

TASK 2: INTERACTIVE TASK

Give students calculators and show them the counting on/constant function. Provide students with a starting number, and then have them count on by 3s. Vary the starting number according to students' abilities. Have students record the number sequence on a sheet of paper. Take note of how students are recording their sequences, e.g. across or down the page, or with the addition included each time. These could be shared later. Give students another starting number and have them try to work out how to use the calculator to count backwards. Again, have students record their number sequences.

TASK 3: STUDENT BOOK p. 24 *'Number Sequences of 3'*

TEACHING GROUP

You will need: BLM 1 'Hundred Chart', counters, coloured pencils, long rolls of paper

SHADES OF THE HUNDRED CHART

- For students who require support, give them each a copy of BLM 1 'Hundred Chart'. Give them a starting number and have them place counters on the number sequence of 3. Check their sequences and ask students to observe any patterns. Have them colour the sequence on the chart in one colour. Then repeat the activity with a different starting number and have students shade the sequence in a different colour. Ask, 'What similarities/differences can you see between the two patterns?' Repeat.

COUNTING ON NUMBER LINES

- For students who require a challenge, have them create scaled number lines, starting at a 5-digit number and continuing the number sequence of 3. This could be done on a long roll of paper, where one square represents a particular division. Ask, 'What do you think the 100th or 253rd number in the sequence may be? How could we find out?'

REFLECTION

Select from the following to suit your class and their learning outcomes:

- Have students share their long number sequences either from Independent Tasks, Task 1, or the Teaching Group 'Counting on Number Lines'. Ask students about the patterns they observe in the numbers. Ask, 'How could this help us predict the 200th number?'
- Have students share their charts from the Teaching Group 'Shades of the Hundred Chart'. Have students show the coloured patterns and make comments about the links between the sequences.
- Revisit NTO 4.1 'Hundred Chart' and have students continue the number sequence off the chart, i.e. count beyond the end of the chart. Ask, 'How would we know the sequence was correct?'

LESSON PLAN 2

TUNING IN

AFL FOOTBALL

You will need: television image from an AFL game that shows the score

Show students a television image from an AFL game that shows the score. Footage can be recorded from the television or can be found at afl.com.au. Talk about the scoring system with students. Record ideas on the board. Draw out that the total score is found by multiplying the number of goals by 6 and then adding the number of points. For example: 6 goals, 10 points $= 6 \times 6 + 10$

$= 36 + 10$

= 46 points (total score)

WHOLE-CLASS INTRODUCTION

HUNDRED CHART: BY 6S

You will need: NTO 4.1 'Hundred Chart' or a Hundred Chart poster

Display NTO 4.1 'Hundred Chart' or a Hundred Chart poster and have students count by the number sequence of 6, starting at a variety of points. Invite students to comment on any patterns they observe on the chart or in the numbers as they are counting. Ask questions, such as, 'What is the next number? How did you work that out?' Examine links with the 3s number sequence. The activity can be extended by having students explore the number sequence backwards.

INDEPENDENT TASKS

Note: Choose from Tasks 1, 2 or 3.

You will need: an A3 copy of BLM 14 'Y Chart' for each student, Student Book p. 25 'AFL Football Scores', *Excel*

TASK 1: LINKS BETWEEN 3 AND 6

Enlarge BLM 14 'Y Chart' to A3 size and make a copy for each student. Have students look at number sequences of 3 on the left-hand side, number sequences of 6 on the right-hand side and then commonalities between the two sequences in the centre. Encourage students to include relevant images, e.g. a triangle for 3s sequences, and an insect for 6s sequences. Students could also write descriptions.

TASK 2: INTERACTIVE TASK

Have students use *Excel* to create a formula to calculate football scores, e.g. = (6*A3) + B3. (The formula needs to give the same result as: 6 × goals + number of points.) Depending on students' abilities, this may be an open task or it could be closely directed with an explanation of the formula function in *Excel*. Have students trial their formula with a set of real scores from the newspaper.

TASK 3: STUDENT BOOK p. 25 *'AFL Football Scores'*

TEACHING GROUP

You will need: BLM 15 'Calculating Football Scores', calculators

CALCULATING FOOTBALL SCORES

- For students who require support, give them a copy of BLM 15 'Calculating Football Scores' and a calculator. As a group, have students work through the BLM.

GIVEN THE SCORE

- For students who require a challenge, give them total scores for football teams for a round of AFL. Have students find at least three different scenarios of scores for each team, either using their *Excel* formula from Independent Tasks, Task 2, or by listing the number sequences of 6. For example, if Brisbane Lions had a score of 78, possible scenarios could be: 10 goals and 18 points; 13 goals and 0 points; 12 goals and 6 points. Provide students with the real break-up at the end and see who was the closest.

REFLECTION

Select from the following to suit your class and their learning outcomes:

- Revisit NTO 4.1 'Hundred Chart' and draw out similarities between the 3s and 6s number sequences. Challenge students to think beyond the end of the Hundred Chart. Ask, 'What number would be next? How could we check?'
- Invite students to share their Y charts from Independent Tasks, Task 1. Look at similarities and differences between students' work. Display the work.
- Have students share possible scores from the Teaching Group 'Given the Score', and reveal the real score.

LESSON PLAN 3

TUNING IN

BUZZ

As a class, play Buzz where the Buzz number is 9. Students stand in a circle and count by 1s. When they reach a multiple of 9, they say 'Buzz!' If someone misses saying 'Buzz!' or miscounts, they are out and sit down in their place. The winner is the last one standing. Challenge students to get into the hundreds.

WHOLE-CLASS INTRODUCTION

NUMBER SEQUENCE: 9S

You will need: NTO 4.1 'Hundred Chart' or a Hundred Chart poster

Have students complete the number sequence of 9s using NTO 4.1 'Hundred Chart' or a Hundred Chart poster. List the number sequence down the board and examine the digits. Ask, 'What do you notice?' (digits add to 9, tens column increases by 1, ones column decreases by 1 and so on) Have students begin at a different starting number. Do their observations hold true? Have students discuss and examine a number of starting points.

INDEPENDENT TASKS

Note: Choose from Tasks 1, 2 or 3.

You will need: BLM 10 'Think Board', BLM 1 'Hundred Chart', modelling equipment such as Unifix blocks, calculators, *Excel*, Student Book p. 26 'Number Sequences of 9'

TASK 1: THINK BOARD 3S, 6S AND 9S

Provide pairs of students with an enlarged copy of BLM 10 'Think Board'. Have students complete the four quadrants, trying to draw links between the number sequences of 3s, 6s and 9s. For the first quadrant, look at the Hundred Chart. For the second quadrant, use number lines; for the third quadrant, use modelling equipment or drawings; and for the final quadrant, write comments.

TASK 2: INTERACTIVE TASK

In *Excel,* have students list down the page the number sequence of 9s between 0 and 100. Next to it, have them list the number sequence of 6s and then 3s. Have students colour or highlight any links between the three columns. Then have them use the patterns and links to extend the numbers to at least 200.

TASK 3: STUDENT BOOK p. 26 ***'Number Sequences of 9'***

TEACHING GROUP

You will need: BLM 1 'Hundred Chart' (two copies for each student), coloured pencils, NTO 4.4 'Number Line: Patterns'

HUNDRED CHART TWO WAYS

- For students who require support, provide copies of BLM 1 'Hundred Chart'. Have them use different-coloured pencils to shade the 9s number sequence, then the 6s and the 3s. Have students list all of the common numbers between two or all three sets of numbers. Give students another copy of BLM 1

'Hundred Chart', and have them repeat the activity for a different starting number. Have them compare the coloured patterns and the list of numbers. Ask, 'What can you see? What is the same? What is different?'

BEYOND AND BELOW 1000

- For students who require a challenge, have them explore the number sequences of 3, 6 and 9 above 1000. Ask, 'Are the number sequences the same as for smaller numbers? Are they different?' To challenge students further, have them consider working backwards and into negative numbers. Ask, 'What happens now?' Students could be scaffolded with NTO 4.4 'Number Line: Patterns'.

REFLECTION

Select from the following to suit your class and their learning outcomes:

- Have students share their think boards from Independent Tasks, Task 1, and make comparisons between similarities and differences between the boards. Display the think boards.
- Have students write a comment, either in *Excel* or in *Word*, about the links they observe and patterns with the number sequences.
- Have students share their charts from the Teaching Group 'Hundred Chart Two Ways'.

Home Tasks

Select from the possible Home Tasks:

- Have students share with parents or carers how they complete number sequences on the calculator.
- Have students ask parents or carers if they know any patterns or 'tricks' with the 3s, 6s or 9s number sequences. Have students share strategies they know or have learnt about. Have students share any ideas from home with the rest of the class.

Assessment

- Have students complete **Student Assessment p. 27**.
- Review with students **Assessment Task Card 4.6**.

During the three lessons:

- Collect the digital materials, such as the *Excel* patterns from Lesson Plan 3, Independent Tasks, Task 2, and add to students' digital portfolios.
- Collect students' think boards from Lesson Plan 3, Independent Tasks, Task 1, and make note of any areas students may have had difficulty with. Add this to students' portfolios as a sample of work.
- Make note of the length and detail of the number sequences devised on rolls of paper in Lesson Plan 1. Note those students who were extended and those who experienced difficulty.

Recommendations for Future Learning

Specific to Student Assessment p. 27; if the student is experiencing some difficulty:

Q 1 & 4 Use NTO 4.1 'Hundred Chart' to explore the number sequences, the overlaps and the patterns.

Q 2 & 4 Use NTO 4.4 'Number Line: Patterns' to explore number sequences from different starting points.

Q 3 Revisit the links between the 3s and 6s number sequences, either through NTO 4.1 'Hundred Chart' or NTO 4.4 'Number Line: Patterns'.

If the student has not achieved the recommended skills for this unit:

1. See **Assessment Task Card 4.6** for specific recommendations.
2. Have the student practise arranging cards in the correct numeric order and then move to building number lines. The cards could be made from BLM 1 'Hundred Chart'.
3. Have the student work with calculators and pens and paper to build the number sequences, and then transfer them to a number line to give a visual representation.
4. Review *Nelson Maths: Australian Curriculum NSW Year 3* Unit 22.

If the student has achieved the recommended skills and these skills are firmly established, consider:

1. Having the student complete 'Counting with Whole Numbers' from *Nelson Maths Building Mental Strategies Big Book 4*, pp. 4–5.
2. Moving forward to *Nelson Maths: Australian Curriculum NSW Year 5* Unit 3.
3. Extending the student in any of the listed activities by using larger numbers, and working both forwards and backwards.

Number Sequences: 4s, 8s and 7s

Number and Algebra
Multiplication and division MA2-6NA uses mental and informal written strategies for multiplication and division

doubling, multiple, number sequence, patterns, place value

LESSON PLAN 1

TUNING IN

FIZZ BUZZ

Students stand in a circle and count by 1s. When a multiple of 2 is reached, the student says 'Buzz!'. When a multiple of 3 is reached, the student says 'Fizz!', and when a multiple of 2 and 3 is reached, e.g. 6, the student says 'Fizz buzz!'. The rest of the rules are the same as Buzz (see Unit 1, Lesson Plan 3, p. 22).

WHOLE-CLASS INTRODUCTION

2S AND 4S NUMBER SEQUENCES

You will need: NTO 4.1 'Hundred Chart'

Display NTO 4.1 'Hundred Chart'. Have students complete number sequences of 2 and discuss patterns. Leave the display on the IWB and have students complete number sequences of 4. Look at commonalities between the two sequences, drawing out the idea of doubling. Have students examine a number of starting points to see if the pattern is consistent.

INDEPENDENT TASKS

Note: Choose from Tasks 1, 2 or 3.

You will need: a range of modelling equipment (e.g. MAB, Unifix blocks, counters), a digital camera, LO: *L2317 'Number trains'*, Student Book p. 28 'Number Sequences of 4'

TASK 1: DOUBLING

Give pairs of students a range of modelling equipment. Have them create and represent a number sequence counting by 2. Then have them double the sequence to complete the number sequence of 4. Have students take digital photos of their work, then write the number sequences and add comments about the development of their work. Note: include students' names in the photos to make identification easier.

TASK 2: INTERACTIVE TASK

Have students complete LO: *L2317 'Number trains'* in a different language to examine number sequences. Languages include French, Greek, Chinese, German, Italian, Indonesian and Japanese.

TASK 3: STUDENT BOOK p. 28 *'Number Sequences of 4'*

TEACHING GROUP

You will need: BLM 1 'Hundred Chart', counters, coloured pencils

FOURS AND THE HUNDRED CHART

- For students who require support, give them each a copy of BLM 1 'Hundred Chart'. Give students a starting number and have them place counters on the number sequence of 2. Check their sequences and ask students to observe any patterns. Have them colour the sequence on the chart in one colour. Then give the same starting number and have students shade the sequence of 4 in a different colour. Ask, 'What similarities/differences can you see between the two patterns?' Have them then investigate the 2s and 4s number sequences, using counters and a different starting number. Ask, 'Does the doubling always work?' If necessary, clarify the term 'doubling'.

TWO + TWO = FOUR

- For students who require a challenge, pose the following problem: How many solutions can you find to this alphanumeric equation? Each letter stands for a different number. There are seven different solutions, e.g. 734 + 734 = 1468.

```
   T W O
 + T W O
 -------
 F O U R
```

REFLECTION

Select from the following to suit your class and their learning outcomes:

- Have students replay Fizz Buzz from 'Tuning In'. See if students are improving with time and accuracy.
- Have students share their photos of the modelled number sequences from Independent Tasks, Task 1. Look for similarities and differences in the representations. Ask students to share comments about why they structured their work in particular ways.
- Have students share their answers to the TWO + TWO problem. Invite them to show their answers on the board. Ask them to share the processes they used to solve the problem.

LESSON PLAN 2

TUNING IN

CARD PATTERNS

You will need: BLM 16 'Number Cards: 4s and 8s'

Copy and cut out the number cards from BLM 16 'Number Cards: 4s and 8s'. Give one card to each student and have them use the number as a starting point to create a number sequence of 4s, then 8s.

WHOLE-CLASS INTRODUCTION

NUMBER LINE: 2S, 4S AND 8S

You will need: NTO 4.4 'Number Line: Patterns'

Display NTO 4.4 'Number Line: Patterns' with the numbers showing and have students count by 2s from different starting points. Invite students to comment on any patterns they may observe on the chart or in the numbers as they are counting. Ask questions, such as, 'What is the next number? How did you work that out?' Repeat with the 4s number sequence and then add the 8s number sequence. Discuss patterns, commonalities and doubling.

INDEPENDENT TASKS

Note: Choose from Tasks 1, 2 or 3.

You will need: lined paper, *PowerPoint*, Student Book p. 29 'Number Sequence Patterns'

TASK 1: LINKS BETWEEN 2S, 4S AND 8S NUMBER SEQUENCES

Give students a lined sheet of paper. Have students list the number sequences of 2, 4 and 8 down the page, with each number on a separate line, up to at least 100 (depending on students' abilities). Then have them look for common numbers between the lists. Students may wish to link these with colour or lines. Then at the bottom of the page, have students write comments about what they found. This activity could be extended by having students repeat the process from a different starting number.

TASK 2: INTERACTIVE TASK

Have students use *PowerPoint* to create a display of the links between the number sequences of 2s, 4s and 8s. Present this as an open-ended task, allowing students to be creative, but reminding them the mathematics is important.

TASK 3: STUDENT BOOK p. 29 *'Number Sequence Patterns'*

Note: BLM 37 'Circles Template' could be used with this activity.

TEACHING GROUP

You will need: BLM 16 'Number Cards: 4s and 8s', coloured pencils

ORDERING CARDS

- For students who require support, give them a copy of BLM 16 'Number Cards: 4s and 8s'. Have students colour the 8s number sequence in one colour and the rest of the cards in another colour. They cut out the cards, then order the 8s number sequence correctly. Ask, 'How did you know that card went there?' Then have them mix up the cards and correctly sequence the 4s number sequence. Ask, 'What do you notice about the colour of the cards?'

OCTAL NUMBER SYSTEM

- For students who require a challenge, have them complete this investigation of numbers in base 8. Say, 'We usually count: 1, 2, 3, 4, 5, ... and so on. In base 8, the numbers 0 to 7 are used and then re-used to represent all other numbers. For example: "one-zero" (written as 10) means 8, "one-one" (written as 11) means 9, and so on. "Two-zero" (written as 20) means two 8s or 16.' Provide a list of numbers for students to express in base 8. This activity could be extended by having students investigate the use of an octal number system.

REFLECTION

Select from the following to suit your class and their learning outcomes:

- Invite students to share their *PowerPoint* presentations from Independent Tasks, Task 2. Ask them to explain why they organised their materials as they did. Look for the mathematics, as well as the design elements.
- Have students share the links they found between the number sequences in Independent Tasks, Task 1. Draw out the concepts of doubling and doubling twice.
- Have students share their investigation from the Teaching Group 'Octal Number System'. Have them explain the system and challenge other students to work out the particular numbers.

LESSON PLAN 3

TUNING IN

HOT POTATO

You will need: a bean bag or a soft ball

As a class, stand in a large circle. Have students throw a bean bag or soft ball to each other in a random pattern. Nominate a number sequence. As each person catches the object, he or she needs to say the next number in the sequence. Start with some simple ones, e.g. 2s and 10s, and see how far students can go. Then move to the 7s number sequence. Discuss some strategies that may help, e.g. add 5 and then 2 more.

WHOLE-CLASS INTRODUCTION

NUMBER SEQUENCES USING CALENDARS

You will need: a calendar on an IWB or a printed calendar

Show students a calendar. Examine the number of days in a week, then the number of weeks in a month. Find the 7th of the month, and follow the number sequence throughout the month. Extend to continue the number sequence through a number of months.

INDEPENDENT TASKS

Note: Choose from Tasks 1, 2 or 3.

You will need: poster paper, *PowerPoint* or *Kid Pix*, Student Book p. 30 'Number Sequences of 7'

TASK 1: 7S POSTER

Give poster paper to pairs of students. Have students list the 7s number sequence down the page to at least 100. Then have students look for patterns, or write strategies to help them continue the number sequence, e.g. + 5 and add 2 to find the next number.

TASK 2: INTERACTIVE TASK

Have students use *PowerPoint* or *Kid Pix* to produce the 7s number sequence, using pictures based on 7. For example: they may start with 7 days of the week, then 14, then 21 and so on. Have students develop a strategy for creating their work, e.g. working in arrays.

TASK 3: STUDENT BOOK p. 30 ***'Number Sequences of 7'***

TEACHING GROUP

You will need: hands-on modelling equipment (e.g. Unifix blocks, counters), mini whiteboards or chalk boards, a digital camera

SHOWING WITH EQUIPMENT

- For students who require support, have groups use modelling equipment to show the 7s number sequence. Have students build the equipment in arrays to link to the calendar activity. This could be completed on mini whiteboards or chalk boards, so students can write the numbers underneath. Have students complete different elements of the number sequence, e.g. one student completes 14, another 21 and so on. Once completed, take photos of the models, then take the equipment away, collect and redistribute the boards and have students organise themselves into the 7s number sequence. Students could also be shown the photos that need to be correctly ordered.

NUMBER OF DAYS

- For students who require a challenge, have them find out how many days since their last birthday, and how many days until their next. Ask them to compare answers with a partner. Then have students write up their method for solving the problem as a description.

REFLECTION

Select from the following to suit your class and their learning outcomes:

- Have students play 'Hot Potato' again, working on the 7s number sequence. Challenge students to reach 100.
- Have students share their *PowerPoint* presentations from Independent Tasks, Task 2, explaining the development of their work, and why they selected the pictures. Have different groups share different starting points.
- Have students share their photos from the Teaching Group 'Showing with Equipment'. Ask questions, such as 'Why are the counters in rows?' Draw out any patterns.

Home Tasks

Select from the possible Home Tasks:

- If possible, have students share digital work with parents or carers, either electronically or in printed form.
- Look for pictures in magazines and newspapers that represent the 4s, 8s or 7s number sequences. Have students bring at least five images to class. These could be used to create a display in the classroom.

Assessment

- Have students complete **Student Assessment p. 31**.
- Review with students **Assessment Task Card 4.7**.

During the three lessons:

- Collect the digital materials, such as the *PowerPoint* presentations and digital photos from Lesson Plan 3, to add to student's digital portfolios.
- Collect students' work showing links between 2s, 4s and 8s number sequences from Lesson Plan 2, Independent Tasks, Task 1, to illustrate their depth of understanding of the number sequences.
- Make note of students completing extension activities and the levels they are achieving.

Recommendations for Future Learning

Specific to Student Assessment p. 31; if the student is experiencing some difficulty:

Q 1 Use NTO 4.4 'Number Line: Patterns' to explore the different number sequences.

Q 2 & 4 Use NTO 4.1 'Hundred Chart' to explore the number sequences, the overlaps and the patterns.

Q 3 Revisit the links between the 4s and 8s number sequences, either through NTO 4.1 'Hundred Chart' or NTO 4.4 'Number Line: Patterns'.

Q 5 Revisit the 7s number sequence using either NTO 4.1 'Hundred Chart' or NTO 4.4 'Number Line: Patterns'. Discuss different strategies to help with finding the next numbers in the sequence.

If the student has not achieved the recommended skills for this unit:

1. See **Assessment Task Card 4.7** for specific recommendations.
2. Have the student practise arranging cards (e.g. from BLM 16 'Number Cards: 4s and 8s') in the correct numeric order and then move to building number lines.
3. Have the student work with a calculator and pen and paper to build the number sequences, and then transfer to a number line to give the visual representation.
4. Have the student continue to work with the Hundred Charts to build understanding of the patterns of the number sequences.
5. Review the calendar activity patterns with students.
6. Review *Nelson Maths: Australian Curriculum NSW Year 3* Unit 22.

If the student has achieved the recommended skills and these skills are firmly established, consider:

1. Having the student complete 'Counting with Whole Numbers' from *Nelson Maths Building Mental Strategies Big Book 4*, pp. 4–5.
2. Moving forward to *Nelson Maths: Australian Curriculum NSW Year 5* Unit 3.
3. Extending the student in any of the listed activities by using larger numbers, and working both forwards and backwards.

Unit 8 Regular Shapes

Measurement and Geometry

Two-dimensional space MA2-15MG manipulates, identifies and sketches two-dimensional shapes, including special quadrilaterals, and describes their features

ML area, centimetres, formal, informal, irregular, regular, shape, units

LESSON PLAN 1

TUNING IN

CLOCKS

You will need: shape blocks

Give groups of students a pile of blocks to sort by criteria they determine. Have groups share their sorting criteria and the different groupings they have created.

WHOLE-CLASS INTRODUCTION

REGULAR SHAPES

You will need: poster paper

Write the term 'shapes' on the board and have students brainstorm what it means. Then write the term 'regular shapes'. Ask, 'What makes a regular shape?' (A regular shape is a polygon where all sides and all angles are equal.) Have students come up with criteria. Record the criteria on poster paper and display.

INDEPENDENT TASKS

Note: Choose from Tasks 1, 2 or 3.

You will need: an assortment of regular shapes, blank cards, LO: *L10736 'Shape sorter: wand tool'*, Student Book p. 32 'Regular Shapes'

TASK 1: WHAT AM I?

Give students an assortment of regular shapes. Have students write 'Who am I?' cards for individual shapes, focusing on features such as whether all angles are equal, the number of equal sides and so on. Have students include the answer on the back of the card. Have them make a number of cards and then test them on a friend.

TASK 2: INTERACTIVE TASK

Have students work independently on computers, using LO: *L10736 'Shape sorter: wand tool'* to examine examples of single shapes. They use the wand tool to compare the sides, right angles and lines of symmetry for each shape, and to work out its features. There are more Learning Objects in this series, which will cater for different learning abilities.

TASK 3: STUDENT BOOK p. 32 *'Regular Shapes'*

TEACHING GROUP

You will need: LO: *L3547 'Tessellations'*, rulers, protractors

ELECTRONIC TESSELLATIONS

- For students who require support, have them work independently on computers, using LO: *L3547 'Tessellations'* to create tessellations with coloured shapes.

CREATING SHAPES

- For students who require a challenge, have them practise drawing exact regular shapes, building on the features they identify as important. Give students rulers and protractors to help with accuracy.

REFLECTION

Select from the following to suit your class and their learning outcomes:

- Have students share their cards from Independent Tasks, Task 1, reading out the description to the class and having the group identify the shapes.

- Have students share their electronic tessellations from the Teaching Group 'Electronic Tessellations'. Ask questions, such as 'Why did you select that shape? How did you know to place that shape there?'
- Have students share their accurate drawings from the Teaching Group 'Creating Shapes'. Ask them to reflect on the difficulty of the task, and how they used the tools, such as the ruler, to help. Display their work.

LESSON PLAN 2

TUNING IN

DRAWING SHAPES

Ask students to draw a regular six-sided shape and an irregular six-sided shape. Have students share their drawings. Ask questions, such as 'Why is this shape regular? How do we know this one is irregular?'

WHOLE-CLASS INTRODUCTION

Go over some of the features of regular shapes, i.e. a polygon where all sides and all angles are equal. Extend to examine what makes a shape irregular. Have students draw a number of different irregular shapes on the board.

INDEPENDENT TASKS

Note: Choose from Tasks 1, 2 or 3.

You will need: BLM 17 '1 cm Grid Paper', an assortment of regular shapes (e.g. blocks), *Word*, Student Book p. 33 'Regular or Irregular?'

TASK 1: FINDING THE AREA

Give students copies of BLM 17 '1 cm Grid Paper' and an assortment of regular shapes. Ask, 'How could we find the area of the shapes?' Guide students towards placing the shape on the grid paper and tracing around it, then counting squares to find the area. Have students write the area (in cm^2) in the centre of the shape. Have them measure the area of three different regular shapes. Then ask students to find three irregular shapes in the classroom and repeat the process.

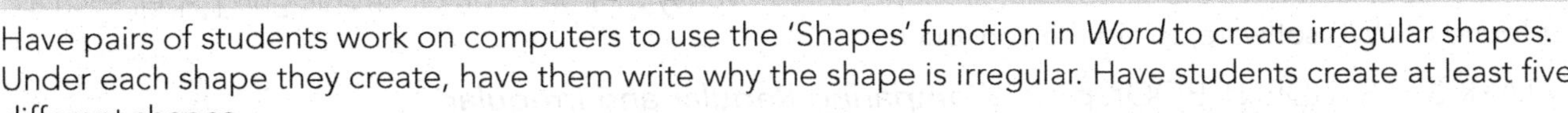

TASK 2: INTERACTIVE TASK

Have pairs of students work on computers to use the 'Shapes' function in *Word* to create irregular shapes. Under each shape they create, have them write why the shape is irregular. Have students create at least five different shapes.

TASK 3: STUDENT BOOK p. 33 *'Regular or Irregular?'*

TEACHING GROUP

You will need: BLM 18 'Regular and Irregular Shapes', scissors, a sheet of paper, glue, BLM 17 '1 cm Grid Paper', internet access

SORTING REGULAR AND IRREGULAR

- For students who require support, give them copies of BLM 18 'Regular and Irregular Shapes' to cut up into cards. Have students sort the shapes into two piles, regular and irregular. Students then identify the features of each set. The cards could be pasted on a sheet of paper under the headings 'Regular' and 'Irregular', with students writing the features of each group in the relevant section. This activity could be extended by having students find the areas of the shapes, placing them on BLM 17 '1 cm Grid Paper' and tracing around them to find the area (note: students would need another copy of BLM 18).

DESIGN WITH SHAPES

- For students who require a challenge, have them play with a design tool such as *Sketchup™*. Allow students to explore what they can design and create.

REFLECTION

Select from the following to suit your class and their learning outcomes:

- Have students write a reflection on the differences in finding the area of regular and irregular shapes in Independent Tasks, Task 1. Ask, 'Which was easier to measure – this shape or this shape? Why?'
- Invite students to share their electronic creations of irregular shapes from Independent Tasks, Task 2. Have students explain how they created certain aspects of the shapes.
- Invite students to share their *SketchUp™* designs from the Teaching Group 'Design with Shapes'. Ask, 'What did you discover about the program?'

TUNING IN

MEASURING AREAS OF REGULAR AND IRREGULAR SHAPES

You will need: a regular shape and an irregular shape (these could be sourced from BLM 18 'Regular and Irregular Shapes')

Show students a regular and an irregular shape. Ask, 'How could we compare the areas of these two shapes?' Have students make suggestions and try out ideas. Guide students towards measuring the respective areas on grid paper, and then comparing.

WHOLE-CLASS INTRODUCTION

MEASURING AREAS OF REGULAR AND IRREGULAR SHAPES

You will need: a grid to project onto the board

Project a large grid onto the whiteboard or IWB. (Note: all IWB software comes with a grid option or these can also be found on the internet.) Ask a student to draw a regular shape on the projection. Have another student work out the area of the shape. Revisit the most effective methods of doing this. Repeat with an irregular shape. Ask, 'What are the units of the area? Which shape has the greatest area? How do we know?' Repeat the activity with different shapes.

INDEPENDENT TASKS

Note: Choose from Tasks 1, 2 or 3.

You will need: BLM 17 '1 cm Grid Paper', *Word*, Student Book p. 34 'Comparing Regular and Irregular'

TASK 1: COMPARING AREA OF SHAPES

Give each student a copy of BLM 17 '1 cm Grid Paper'. Give pairs of students the challenge of finding at least two shapes – one regular and one irregular – that have the same area. Students may select anything in the classroom – suggest that they could create irregular shapes from regular shapes. Have students find as many different pairs as they can and record their findings.

TASK 2: INTERACTIVE TASK

Have students work individually on computers to use the 'Shapes' function in *Word* to create regular and irregular shapes that have the same area. Students could overlay their work on electronic grid paper, such as that found on IWB software, to aid the construction.

TASK 3: STUDENT BOOK p. 34 *'Comparing Regular and Irregular'*

TEACHING GROUP

You will need: BLM 17 '1 cm Grid Paper', flat blocks, LO: *L3547 'Tessellations'*

COMPARING AREAS OF BLOCKS

- For students who require support, give them a copy of BLM 17 '1 cm Grid Paper' and some flat blocks. Have students look at the blocks and estimate two that may be of equal area. Have them record their estimation. Then have students place the blocks on the grid paper, trace around them and then find the areas. Ensure that students count the squares correctly and talk to them about how to count the partial squares. Have them check their prediction. Repeat the activity.

CREATING TESSELLATIONS

- For students who require a challenge, have them work independently on computers, using LO: *L3547 'Tessellations'* to create tessellations with coloured shapes. Have students complete a tessellation, then identify which tiles are regular and which are irregular. Have students make a comment about whether the regularity (or otherwise) of the tiles had an impact on creating tessellating patterns.

REFLECTION

Select from the following to suit your class and their learning outcomes:

- Have students share their equal area shapes from Independent Tasks, Task 1. Ask, 'How did you find the two shapes? Did you have a strategy? How did you check that the areas were the same?'
- Have students share the equal area shapes from Independent Tasks, Task 2. Ask, 'How did you know the areas were the same? How did you check? Which shape is regular and why?' Students could show their work electronically.
- Invite students from the Teaching Group 'Creating Tessellations' to share their tessellation patterns. Ask them to comment on whether the regularity (or otherwise) of the tiles had an impact on creating tessellating patterns.

Home Tasks

Select from the possible Home Tasks:

- Have students conduct a shape hunt at home, looking for regular and irregular shapes. Have them list five things from the kitchen that have a regular shape and five that have an irregular shape. Have students bring the items (if possible) or their list to school to share.
- Give each student a copy of BLM 17 '1 cm Grid Paper'. Have them find three items at home (regular or irregular) that have an area of 10 cm^2. Have students trace around the items and label.

Assessment

- Have students complete **Student Assessment p. 35**.
- Review with students **Assessment Task Card 4.8**.

During the three lessons:

- Collect copies of students' electronic designs, such as the equal area shapes in Lesson Plan 3, Independent Tasks, Task 2, to add to students' digital portfolios.
- Collect students' written reflections, such as the reflection on finding the area of regular and irregular shapes in Lesson Plan 2.
- Collect the home tasks of the shape hunt and finding the area of shapes, illustrating students' understanding of the concepts.

Recommendations for Future Learning

Specific to Student Assessment p. 35; if the student is experiencing some difficulty:

Q 1–2 Review the differences between shapes. Have the student sort shapes into groups and then identify properties of particular groups. Repeat, separating regular and irregular shapes.

Q 3 Have the student practise tracing shapes and counting squares. Show the student strategies for recording the number of squares, such as numbering each individual square or colouring in and keeping a tally. Examine how to count partial squares. Talk about the importance of units.

If the student has not achieved the recommended skills for this unit:

1. See **Assessment Task Card 4.8** for specific recommendations.
2. Work with simple shapes, such as squares, rectangles and triangles. Consolidate ideas with these shapes before moving forward.
3. Practise counting squares to determine areas.
4. When making area comparisons, have the student cut out the shapes and try to overlay them to determine which one is larger.
5. Review *Nelson Maths: Australian Curriculum NSW Year 3* Unit 29.

If the student has achieved the recommended skills and these skills are firmly established, consider:

1. Having the student investigate circles and how they fit into the classification. They may wish to use references.
2. Moving forward to *Nelson Maths: Australian Curriculum NSW Year 5* Unit 8.
3. Having the student complete problems that require multiple steps to solve, e.g. nominate an area of x amount, then create both an irregular and a regular shape of this area.

2D Shapes and 3D Objects

Measurement and Geometry

Three-dimensional space MA2-14MG makes, compares, sketches and names three-dimensional objects, including prisms, pyramids, cylinders, cones and spheres, and describes their features

Two-dimensional space MA2-15MG manipulates, identifies and sketches two-dimensional shapes, including special quadrilaterals, and describes their features

common shapes, composite shape, regular, shape, two-dimensional, three-dimensional, three-dimensional objects

LESSON PLAN 1

TUNING IN

SORTING SHAPES

You will need: 2D shape blocks, large sheets of paper

Give groups of students a pile of blocks to sort by criteria they determine. Have students trace the shapes on paper and create the groupings with a written description. Have students share their work.

WHOLE-CLASS INTRODUCTION

WHAT DEFINES A SHAPE?

You will need: 2D shape blocks, NTO 4.9 '2D Shapes and 3D Objects'

Write the terms 'face', 'edge' and 'vertices' on the board. Have students brainstorm what these aspects are in terms of 2D shapes. Have students classify shapes using an aspect such as the number of edges.
NTO 4.9 '2D Shapes and 3D Objects' could be used to remind students of different shapes and their names.

INDEPENDENT TASKS

Note: Choose from Tasks 1, 2 or 3.

You will need: LO: *L1071 'Shape overlays: picture studio'*, NTO 4.9 '2D Shapes and 3D Objects', Student Book p. 36 '2D Shapes'

TASK 1: DRAW MY SHAPE

Give students a set of verbal instructions to draw a common shape, such as a square. Include descriptions of the number of edges and vertices, if the sides are equal length and where these are orientated. Give instructions for a different shape, such as a triangle. Then have students write three different sets of instructions for drawing a shape. Support students by showing NTO 4.9 '2D Shapes and 3D Objects'.

TASK 2: INTERACTIVE TASK

Have students work independently on computers, using LO: *L1071 'Shape overlays: picture studio'* to position two simple shapes to form an overlap, then cut out that new shape. They then go on to make pictures.

TASK 3: STUDENT BOOK p. 36 *'2D Shapes'*

Note: support students by showing NTO 4.9 '2D Shapes and 3D Objects'.

TEACHING GROUP

You will need: 2D shape blocks

NAMING SHAPES

- For students who require support, have them trace a 2D shape on a sheet of paper. Students record the name of the shape and list the attributes, e.g. the number of edges. Have students use different colours to highlight different aspects of the shape. Students repeat with a number of shapes, creating a reference list to work with.

CREATING SHAPES

- For students who require a challenge, give them a number of 2D shapes and have them create a composite shape on a sheet of paper. Have them trace around the blocks as a record, then write a set of instructions for creating this 'new' shape.

REFLECTION

Select from the following to suit your class and their learning outcomes:

- Have students share their instructions from Independent Tasks, Task 1. Have a student read out the instructions and have the other students draw the relevant shape.
- Have students match the names of the shapes and the shapes on NTO 4.9 '2D Shapes and 3D Objects'.
- Have students from the Teaching Group 'Creating Shapes' read out their 'new' shape instructions and have the other students in the class make the shape with blocks. Ask, 'Were they right? Were the instructions clear?'

LESSON PLAN 

TUNING IN

ELECTRONIC TANGRAM

You will need: a tangram program

On the IWB, show a tangram program such as: http://nlvm.usu.edu/en/nav/frames_asid_289_g_2_t_3.html?open=activities&from=category_g_2_t_3.html. Have students attempt the puzzle.

WHOLE-CLASS INTRODUCTION

You will need: BLM 19 'Tangram Instructions', sheets of paper, rulers, scissors, envelopes

Give students some paper. Have students follow your verbal instructions (you can use BLM 19 'Tangram Instructions') to make a tangram. See how successful they were. Note: students should store their tangrams in labelled envelopes.

INDEPENDENT TASKS

Note: Choose from Tasks 1, 2 or 3.

You will need: BLM 20 'Tangram Shapes to Make', BLM 21 'Tangram Template', a digital camera, internet access, electronic tangrams such as those on the *My First Tangrams* app or http://nlvm.usu.edu/en/nav/frames_asid_289_g_2_t_3.html?open=activities&from=category_g_2_t_3.html, Student Book p. 37 'Features of Tangrams'

TASK 1: USING TANGRAMS

Give students copies of BLM 20 'Tangram Shapes to Make'. Have students use the tangram they made in the Whole-Class Introduction to try to work out the shapes on the BLM. Make sure students record their answer by drawing or taking a digital photo. Note: BLM 21 'Tangram Template' could also be used for this activity.

TASK 2: INTERACTIVE TASK

Using websites such as http://nlvm.usu.edu/en/nav/frames_asid_289_g_2_t_3.html?open=activities&from=category_g_2_t_3.html or an app such as *My First Tangrams*, have students explore tangrams on the IWB or individually on computers.

TASK 3: STUDENT BOOK p. 37 *'Features of Tangrams'*

TEACHING GROUP

You will need: BLM 22 'Making a Tangram with a Grid', BLM 23 'More Tangrams!', BLM 17 '1 cm Grid Paper', rulers, scissors, *Word*

USING A GRID TO MAKE A TANGRAM

- For students who require support, give them each a copy of BLM 22 'Making a Tangram with a Grid'. Have students use rulers to copy the tangram from the top of the page to the grid on the bottom. Discuss how to line up the ruler, e.g. with the corners of squares on the grid. Students could cut out their tangrams when finished, or they could try to recreate the tangrams on BLM 23 'More Tangrams!' using BLM 17 '1 cm Grid Paper'.

ELECTRONIC TANGRAM

- For students who require a challenge, have them use the 'Shapes' function in *Word* to create their own tangram. To challenge students further, give them a copy of BLM 23 'More Tangrams!' and have them create the alternative tangrams electronically. Once they have completed their on-screen tangrams, they could then create and save some shapes/pictures using the tangram.

REFLECTION

Select from the following to suit your class and their learning outcomes:

- Have students sit in a circle with the tangrams they made in the Whole-Class Introduction. Ask, 'What was difficult about making a tangram just from verbal instructions? What part did you find the hardest?' Examine which students achieved the task and who struggled.
- Invite students to share which shapes they made in Independent Tasks, Task 1. Ask, 'Did anyone manage to complete them all?' Ask students to share their recording (drawn or digital).

- Have students from the Teaching Group 'Using a Grid to Make a Tangam' describe how they used the grids to make the tangrams. Ask, 'Was it easier to use the grid or to follow my instructions? Why?'
- Invite students to share the tangrams and shapes from the Teaching Group 'Electronic Tangrams', describing how they created the tangram using the 'Shapes' function.

LESSON PLAN 3

TUNING IN

A SQUARE AND A CUBE

You will need: a cube, such as an MAB 1000 block

Ask a student to draw a square on the board. Show the cube and ask students to reflect on the similarities and differences between a square and a cube. Students then share their ideas. Lead (if necessary) the discussion to the conclusion that a square can be seen in two dimensions and a cube in three dimensions.

WHOLE-CLASS INTRODUCTION

THREE-DIMENSIONAL DRAWINGS

Remind students of the cube and the drawing of the square that they looked at in the Tuning In session. Ask students why a square is easier to draw than a cube. Reassure them that making a two-dimensional representaion of a three-dimensional object is a skill that even some adults have difficulty with. Ask for a volunteer to draw a cube on the board and to explain how they did it. Provide lots of scrap paper and invite students to sketch a cube, then invite students to share how they completed the drawings, to share skills. BLM 32 'Isometric Dot Paper' could also be used for the sketches.

INDEPENDENT TASKS

Note: Choose from Tasks 1, 2 or 3.

You will need: BLM 32 'Isometric Dot Paper', NTO 4.11 'Isometric Dot Paper', Student Book p. 38 'Drawing 3D Objects'

TASK 1: 3D SKETCH

Provide isometric dot paper (BLM 32) and allow students time and opportunity to practise sketching 3D objects. This activity would ideally take place during the work on Student Book p. 38 'Drawing 3D Objects'. Students could be scaffolded by being shown how to overlap shapes and join common vertices to form 3D objects.

TASK 2: INTERACTIVE TASK

Have students work individually on NTO 4.11 'Isometric Dot Paper' to practise digital sketches of 3D objects.

TASK 3: STUDENT BOOK p. 38 *'Drawing 3D Objects'*

TEACHING GROUP

You will need: BLM 32 'Isometric Dot Paper', scrap paper, Student Book p. 38 'Drawing 3D Objects'

PRACTISE, PRACTISE, PRACTISE

- For students who require support, discuss the rectangular prism on p. 38 of the Student Book. Ask them to look at the dot and arrow next to the rectangular prism and to identify on the paper the dot to which the line should go. Repeat until the rectangle for the front face has been drawn. Next draw the diagonal line from the top left vertex, each time identifying the next dot. Allow time and opportunity for lots of trial and error.

SKETCHING FROM A CONCRETE OBJECT

- Students who can easily and successfully copy the 3D objects can be invited to try to draw a 3D object by looking at the object itself rather than a drawing of one. They could begin with one that is on the page and move on to other objects such as a sphere or a hexagonal prism.

REFLECTION

Select from the following to suit your class and their learning outcomes:

- Ask, 'What would be difficult about drawing a sphere?' Have students share drawings of spheres.
- Have students write a reflection on what they feel are the challenges of drawing three-dimensional objects.
- Ask, 'If you draw a cube, how many faces can be seen?' Have students check responses by looking at different examples of 3D drawings of cubes.

Home Tasks

Select from the possible Home Tasks:

- Ask students to identify the most common 3D objects found in their home.
- Have students make a list of 3D objects at home that are *not* rectangular prisms.
- Invite students to have a parent or carer make a freehand sketch of a cube or other 3D object.

Assessment

- Have students complete **Student Assessment p. 39**.
- Review with students **Assessment Task Card 4.9**.

During the three lessons:

- Invite students to let you have their first and final attempts at drawing a cube or other 3D object.
- Collect students' written reflections, such as the reflection on drawing 3D objects in Lesson Plan 3.
- Collect digital materials such as those made in Lesson Plan 3, Independent Tasks, Task 2, as evidence for students' digital portfolios. Students may wish to add a comment about the activity.

Recommendations for Future Learning

Specific to Student Assessment p. 39; if the student is experiencing some difficulty:

Q 1 Review the different common 2D shapes. Revisit their properties and the names of each shape. NTO 4.9 '2D Shapes and 3D Objects' could be used to support recognition.

Q 2 Revisit the idea of a composite shape. Use blocks, or similar materials, to construct and pull apart composite shapes. Examine the common shapes that make up the composite shape, beginning with composite shapes that use only two common shapes.

Q 3 Reassure students that the ability to draw 3D objects is not a skill that is acquired at a specific age.

If the student has not achieved the recommended skills for this unit:

1. See **Assessment Task Card 4.9** for specific recommendations.
2. Continue with the hands-on manipulation of materials. Use blocks, cut-out shapes and tactile materials, such as plastic, foam and felt. Have the student move shapes together and apart.
3. Use a digital camera to make the recording of the work quicker and easier for the student. Include the student's name in the photo for later identification.
4. Present the student with a wide range of experiences with shape-based puzzles, either in digital form or with hands-on materials. Encourage the student to persist, and not to look at answers when the problem solving becomes challenging.
5. Review *Nelson Maths: Australian Curriculum NSW Year 3* Unit 29.

If the student has achieved the recommended skills and these skills are firmly established, consider:

1. Having the student work with more complex composite shapes.
2. Moving forward to *Nelson Maths: Australian Curriculum NSW Year 5* Unit 9.
3. Having the student explore the perimeter and area of composite shapes.
4. Giving the student access to more complex shape-based puzzles, either in digital form or with hands-on materials.

Multiplication Facts (Times Tables)

Number and Algebra

Multiplication and division MA2-6NA uses mental and informal written strategies for multiplication and division

 multiplication, place value, tables

LESSON PLAN 1

TUNING IN

BINGO!

You will need: BLM 3 'Tables Chart 1', BLM 4 'Tables Chart 2'

Have students create a 3 × 3 grid and write in various answers from the 2 and 4 times tables up to 10. Play 'Bingo', having students cross off the answers to tables you read out. For example: read out '4 × 2'; if the student has 8 on their grid, they can cross it off. The winner is the first student with all nine numbers crossed off. Play a number of rounds, using at least two sets of tables. If students are struggling with their tables, they may benefit from having copies of BLM 3 'Tables Chart 1' and BLM 4 'Tables Chart 2'. These might also be useful for the person reading out the questions.

WHOLE-CLASS INTRODUCTION

ARRAYS

Use a grid on the IWB or draw one on the board. Have students explore the different array representations of a number, e.g. 12. Have students try to find all possibilities, e.g. 12 × 1, 1 × 12, 6 × 2, 2 × 6, 4 × 3, 3 × 4. Discuss the difference between, say, 1 × 12 and 12 × 1, as well as the importance of labelling each array.

INDEPENDENT TASKS

Note: Choose from Tasks 1, 2 or 3.

You will need: BLM 17 '1 cm Grid Paper', BLM 3 'Tables Chart 1', BLM 4 'Tables Chart 2', computers, LO: *L106 'The array'*, LO: *L108 'The array: go figure'*, Student Book p. 40 'Times Tables'

TASK 1: ARRAYS

Give students a set of numbers with which to explore different arrays, e.g. 45, 20 and 16. Have students explore all of the variations and them draw and label them on BLM 17 '1 cm Grid Paper'. Students may need to use separate pieces of grid paper for each number. Ask them to write a heading on their work. This activity can be differentiated by providing students with numbers suited to their abilities. For students who require support, provide BLM 3 'Tables Chart 1' and BLM 4 'Tables Chart 2'.

TASK 2: INTERACTIVE TASK

Have students work independently on computers, using LO: *L106 'The array'* and LO: *L108 'The array: go figure'*, depending on their ability. Students use an array-building tool to help solve multiplications by exploring strategies to break up numbers.

TASK 3: STUDENT BOOK p. 40 *'Times Tables'*

TEACHING GROUP

You will need: LO: *L2058 'Pobble arrays: find two factors'*

LO: L2058 'POBBLE ARRAYS: FIND TWO FACTORS'

- For students who require support, have them work independently on computers, using LO: *L2058 'Pobble arrays: find two factors'* to work on their basic multiplication and factor skills, while using the arrays to support their learning.

TABLES PUZZLE

- For students who require a challenge, have them create a puzzle page based on tables up to 10 × 10. Encourage students to be creative, perhaps having a word solution to find. Students may need computers for this activity.

REFLECTION

Select from the following to suit your class and their learning outcomes:

- Have students replay 'Bingo' to practise tables skills.
- Have students share the arrays they developed in Independent Tasks, Task 1. Display the array sets around the room. Have students comment about how they found all of the combinations.
- Print or photocopy the puzzles devised by students in the Teaching Group 'Tables Puzzle' and have others try to solve them.

LESSON PLAN 2

TUNING IN

TABLES PATTERNS

You will need: BLM 3 'Tables Chart 1', BLM 4 'Tables Chart 2', A3 sheets of coloured paper

Copy BLM 3 'Tables Chart 1' and BLM 4 'Tables Chart 2', and cut out the individual sets of tables. Divide the class into groups and give each group a set, e.g. the 8s. Have them paste this in the centre of an A3 sheet of coloured paper. Then have students look at all the patterns they can find in the table set and write these on the paper. Students also include any hints or tricks they may be familiar with, e.g. for the 8s, it is double the 4s.

WHOLE-CLASS INTRODUCTION

9 TIMES TABLES TRICKS

You will need: NTO 4.4 'Number Line: Patterns'

Teach students the nine times tables finger trick (see the top of Student Book p. 41 '9 × Tables'). Have students practise a number of times.

INDEPENDENT TASKS

Note: Choose from Tasks 1, 2 or 3.

You will need: two 10-sided dice for each pair, BLM 24 'Rolling Tables', LO: *L2060 'Arrays: explore factors'*, LO: *L108 'The array: go figure'*, Student Book p. 41 '9 × Tables'

TASK 1: ROLLING TABLES

Have pairs of students play the BLM 24 'Rolling Tables' game.

TASK 2: INTERACTIVE TASK 

Have students work independently on computers, using LO: *L2060 'Arrays: explore factors'* and LO: *L108 'The array: go figure'* to explore factors in an array format. Select the Learning Object according to students' abilities.

TASK 3: STUDENT BOOK p. 41 *'9 × Tables'*

TEACHING GROUP

You will need: a set of cards on which the answers to the nine times tables are written (one answer per card), calculators

ORDERING CARDS

- For students who require support, give them a set of cards with the answers to the nine times tables written on them. Have students place them in the correct order of the answers vertically on a sheet of paper. Next to the answers, have students write the equation. Have students look at patterns in the answers, i.e. the tens digits increase while the ones digits decrease. Take the cards away and challenge students to write the answers to the tables.

EXPLORING PATTERNS

- For students who require a challenge, have them explore how far the 'tricks' from Student Book p. 41, and any other patterns, will work for the nine times tables, e.g. in the answers, the tens digits increase while the ones digits decrease. Ask students to see if there are any other tricks for higher numbers. Students may wish to use calculators in their investigation.

REFLECTION

Select from the following to suit your class and their learning outcomes:

- Invite groups to present their posters from Tuning In. Students may wish to have a few minutes to add any further ideas to their posters.
- Have students practise their nine times tables 'tricks' for some tables you call out. Ask them to comment on which strategy they prefer and why.
- Have students from the Teaching Group 'Exploring Patterns' share their investigations on how far the tricks will work for the nine times tables.

LESSON PLAN 3

TUNING IN

REVERSE BINGO

You will need: BLM 3 'Tables Chart 1', BLM 4 'Tables Chart 2'

Play 'Bingo' as in Tuning In in Lesson Plan 1, but this time, students write the equations in their grids and the answers are read out. The winner is the first student with all nine squares crossed off. Play a number of rounds, using at least two sets of tables. If students are struggling with their tables, they may benefit from having copies of BLM 3 'Tables Chart 1' and BLM 4 'Tables Chart 2'. These might also be useful for the person reading out the answers.

WHOLE-CLASS INTRODUCTION

NUMBER SEQUENCES USING CALENDARS

You will need: NTO 4.1 'Hundred Chart', NTO 4.4 'Number Line: Patterns'

Use NTO 4.1 'Hundred Chart' to explore tables patterns on the Hundred Chart, inviting students to the board for activities such as completing tables patterns, continuing tables patterns or creating and displaying patterns they may be aware of. Work with sets of tables such as 2s, 4s and 8s. Look at the pattern made by the 9s. Also explore representations of tables (via repeated addition) on NTO 4.4 'Number Line: Patterns'.

INDEPENDENT TASKS

Note: Choose from Tasks 1, 2 or 3.

You will need: poster paper, *Word*, Student Book p. 42 'More Tables'

TASK 1: LINKS BETWEEN TABLES

Give pairs of students poster paper. Provide them with sets of tables, e.g. 5s and 10s, and have them create a poster highlighting the links between the tables. Ensure students list the sets of tables and link any commonalities.

TASK 2: INTERACTIVE TASK

Have students select two sets of tables that they need to practise. Students type up the tables on *Word* and create mini posters they can use to practise with. They may wish to include hints and strategies on their posters. For students who require a further challenge, extend the tables sets beyond 10 × 10.

TASK 3: STUDENT BOOK p. 42 ***'More Tables'***

TEACHING GROUP

You will need: blank cards, BLM 3 'Tables Chart 1', BLM 4 'Tables Chart 2', snap-lock bags, cardboard, dice, counters

MY FLASH CARDS

- For students who require support, have them make a set of tables flash cards to practise tables with. Give students blank cards and copies of BLM 3 'Tables Chart 1' and BLM 4 'Tables Chart 2' (circle the tables sets that students should focus on). The cards should have the question on one side and the answer on the other. Allow students to decorate the cards. When finished, have students practise with their cards. Students could keep the cards in named snap-lock bags.

CREATING A GAME

- For students who require a challenge, have them create a board game based on tables. These could be based on games that students are familiar with, e.g. 'Snakes and Ladders'. Provide materials such as cardboard, counters and dice.

REFLECTION

Select from the following to suit your class and their learning outcomes:

- Have students play 'Reverse Bingo' again, challenging them to play without using the BLMs as support materials.
- Have students share the posters they made in Independent Tasks, Task 1. Invite students to share what they found. Look for commonalities and differences between the groups.
- Have students from the Teaching Group 'Creating a Game' share the board games they made. Students could play the games in small groups.

Home Tasks

Select from the possible Home Tasks:

- Have students take home their mini posters from Lesson Plan 3, Independent Tasks, Task 2, and practise their tables.
- Have students teach their parents or carers the nine times tables finger trick.
- Have students who made flash cards or board games in Lesson Plan 3, Teaching Group, take them home to practise with parents or carers.

Assessment

- Have students complete **Student Assessment p. 43**.
- Review with students **Assessment Task Card 4.10**.

During the three lessons:

- Take copies of students' work to collect as evidence of tables work for their portfolios.
- Make note of the tables sets students are struggling with, and those they have mastered.
- Review Student Book pages to investigate areas where students may need support.

Recommendations for Future Learning

Specific to Student Assessment p. 43; if the student is experiencing some difficulty:

Q 1–2 Give the student copies of BLM 3 'Tables Chart 1' and BLM 4 'Tables Chart 2' to practise the sets of tables they are not confident with.

Q 3–4 Revisit tables using materials such as BLM 3 'Tables Chart 1', BLM 4 'Tables Chart 2' and NTO 4.1 'Hundred Charts'.

Q 5 Review tricks to solve nine times tables from Lesson Plan 2, Whole-Class Introduction.

Q 6 Review arrays. Use concrete materials such as counters to aid explanations.

If the student has not achieved the recommended skills for this unit:

1. See **Assessment Task Card 4.10** for specific recommendations.
2. Have the student use BLM 3 'Tables Chart 1' and BLM 4 'Tables Chart 2' to practise tables.
3. Have the student work with cards that have tables equations on them and play simple games to practise automatic recall.
4. Have the student continue to work with Hundred Charts (hard copy or electronic) to build understanding of the patterns of the number sequences.
5. Review *Nelson Maths: Australian Curriculum NSW Year 3* Units 18 and 19.

If the student has achieved the recommended skills and these skills are firmly established, consider:

1. Having the student complete 'Multiplying by 8, 3, 6, 9 and 7' from *Nelson Maths Building Mental Strategies Big Book 4*, pp. 50–59.
2. Moving forward to *Nelson Maths: Australian Curriculum NSW Year 5* Unit 7.
3. Extending the student in any of the listed activities by moving to the 11 and 12 times tables.

Unit 11 Multiplication Facts and Related Division Facts

Number and Algebra

Multiplication and division MA2-6NA uses mental and informal written strategies for multiplication and division

division, fact families, multiplication, place value, tables

LESSON PLAN 1

TUNING IN

TABLES CHALLENGE

You will need: BLM 3 'Tables Chart 1', BLM 4 'Tables Chart 2'

In groups, have students complete a tables challenge. Include questions such as: '4 × 5 = ? What tables equal 24? How many legs do five chickens have?' Play a number of rounds. If students require support, provide BLM 3 'Tables Chart 1' and BLM 4 'Tables Chart 2'.

WHOLE-CLASS INTRODUCTION

MULTIPLICATION GRID

You will need: NTO 4.1 'Hundred Chart'

Display NTO 4.1 'Hundred Chart'. Discuss tables and counting patterns. Introduce the concept of a multiplication grid. Explain that the numbers 1–10 are listed down one side and across the top of the grid, and the numbers inside are the answers to the tables. Start showing students how to construct and work out the numbers in the grid.

INDEPENDENT TASKS

Note: Choose from Tasks 1, 2 or 3.

You will need: poster paper, rulers, LO: *L83 'The multiplier: generate easy multiplications'*, Student Book p. 44 'Using Your Multiplication Grid'

TASK 1: MULTIPLICATION GRID

Give students poster paper and rulers and have them make their own multiplication grid. This activity could be modified by providing part of the table for students who require more support, and it can be extended by having students complete up to the 12 times tables.

TASK 2: INTERACTIVE TASK

Have students work independently on computers, using LO: *L83 'The multiplier: generate easy multiplications'* to practise skills to help them solve more complex multiplication equations.

TASK 3: STUDENT BOOK p. 44 *'Using Your Multiplication Grid'*

TEACHING GROUP

You will need: students' own multiplication grids, overhead film and whiteboard markers or two rulers per student, LO: *L61 'The multiplier: make your own easy multiplications'*, LO: *L82 'The multiplier: make your own hard multiplications'*

MULTIPLICATION GRID

- For students who require support, laminate their multiplication grids (or overlay them with overhead film). Work as a group and pose a tables question, e.g. 4 × 6. Have students use a whiteboard marker to circle the first number down the side (4), then the second number across the top (6), and then follow the lines to meet the answer. Alternatively, students could use two rulers to find the intersection. Practise this with a number of questions. If students need a challenge, give them an answer, e.g. 49, and have them work backwards to find the question.

MAKE YOUR OWN MULTIPLICATIONS

- For students who require a challenge, have them work independently on computers, using *LO: L61 'The multiplier: make your own easy multiplications'* or LO: *L82 'The multiplier: make your own hard multiplications'* to practise skills to help them solve more complex multiplication equations.

REFLECTION

Select from the following to suit your class and their learning outcomes:

- Play some more rounds of the tables challenge from Tuning In, building on questions and scores from the beginning of the lesson. Allow students to use their multiplication grid for these rounds.
- Have students share their multiplication grids from Independent Tasks, Task 1. Have students find the answers to particular equations using their grids.
- As a group, go through the answers to Student Book p. 44 'Using Your Multiplication Grid'. Ensure students are using the grid correctly.

LESSON PLAN 2

TUNING IN

MATCHING PAIRS

You will need: BLM 25 'Matching Equations'

Copy and cut up the cards from BLM 25 'Matching Equations'. Give each student one card, and ask them to find all of the members of the related tables group. This activity could be extended by having students work in silence. When complete, collect the cards, mix and redistribute. Repeat a number of times.

WHOLE-CLASS INTRODUCTION

FACT FAMILES

You will need: BLM 25 'Matching Equations'

Using the cards from BLM 25, attach $4 \times 3 = 12$ and $3 \times 4 = 12$ to the board. Next to them, write all of the other options that create the facts family. Include $12 = 4 \times 3$ and $12 = 3 \times 4$, and expand to include $12 \div 3 = 4$ and $4 = 12 \div 3$ to find all eight options. Repeat with another example, such as 6×2.

INDEPENDENT TASKS

Note: Choose from Tasks 1, 2 or 3.

You will need: two 10-sided dice per pair, LO: *L2059 'Arrays: factor families'*, Student Book p. 45 'Fact Families'

TASK 1: INVESTIGATING FACT FAMILIES

Provide pairs of students with two 10-sided dice. Have students roll the dice, with each dice forming the number for a multiplication table, e.g. 4×6. Have students find all the related multiplication and division equations, and record them on a sheet of paper. When students have found all options, they roll the dice again for another equation. Students who require support could use their multiplication grid from Lesson Plan 1.

TASK 2: INTERACTIVE TASK 

Have students work independently on computers, using LO: *L2059 'Arrays: factor families'* to explore factors in an array format.

TASK 3: STUDENT BOOK p. 45 ***'Fact Families'***

TEACHING GROUP

You will need: an egg carton or muffin tray, calculators

REAL-LIFE ARRAYS

- For students who require support, use a prop such as an egg carton or a muffin tray. Look at the array in the prop. Have students write all of the related multiplication and division equations based on the prop. Have students share their responses. When all possibilities are exhausted, repeat with another prop.

FACT FAMILIES OF LARGE NUMBERS

- For students who require a challenge, have them explore fact families of larger numbers, particularly working into the 12 times tables. Challenge students to find as many related multiplication and division equations as possible. Students could be given larger numbers into the hundreds or thousands, but may need a calculator.

REFLECTION

Select from the following to suit your class and their learning outcomes:

- Invite the groups to share their fact family investigations from Independent Tasks, Task 1. Look for students who had common equations – did they find all of the alternatives?
- Present students with a prop such as the tray from a box of chocolates. Discuss the fact family related to the array in the prop, with students listing all related equations on the board.
- Have students from the Teaching Group 'Fact Families of Large Numbers' share their investigation.

LESSON PLAN 3

TUNING IN

BINGO DIVISION

You will need: BLM 3 'Tables Chart 1', BLM 4 'Tables Chart 2'

Play 'Bingo' as in Unit 10, Lesson Plan 1, Tuning In, but this time it will be based on the division tables rather than multiplication. Students write the answers in their grids and the equations are read out. The winner is the first student with all nine squares crossed off. Play a number of rounds, using at least two sets of tables. If students are struggling with their tables, they may benefit from having copies of BLM 3 'Tables Chart 1' and BLM 4 'Tables Chart 2'. These might also be useful for the person reading out the equations.

WHOLE-CLASS INTRODUCTION

DIVISION WITH ARRAYS

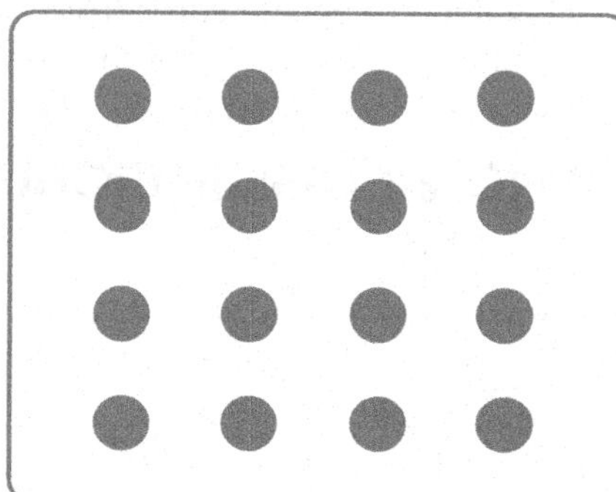

Draw on the board a set of circle arrays, such as those shown at right. Have students identify the total number of circles. Ask different students to come to the board and divide the array into three groups, circling the circles appropriately. There will be more than one answer. Write the relevant multiplication equation, and then division equations. Explore a number of examples.

INDEPENDENT TASKS

Note: Choose from Tasks 1, 2 or 3.

You will need: poster paper, *Kid Pix,* Student Book p. 46 'Division With Arrays'

TASK 1: DIVISION WITH ARRAYS

Give students a number of tables questions. Students draw an array for a question, then explore and show on the array the relevant division equations, thus leading to the fact families. Provide the tables questions according to students' abilities. This activity could be extended by having students create a poster showing their work.

TASK 2: INTERACTIVE TASK

Have pairs of students use computers to create arrays on a program such as *Kid Pix.* Provide tables according to students' abilities. Allow students to be creative, but they need to show the array clearly, and then the separation for the division. Have students also write relevant equations with the work.

TASK 3: STUDENT BOOK p. 46 *'Division With Arrays'*

Note: Students can be encouraged to find equations beyond reversing the multiplication equation. For example, with 6 × 4, students could find 24 ÷ 4, 24 ÷ 6, 24 ÷ 2 and so on.

TEACHING GROUP

You will need: counters or individual whiteboards and markers

ARRAYS WITH COUNTERS

- For students who require support, give them a number of counters and a multiplication equation, according to their ability. Have students create the array with counters. Then have students complete the division equations by moving the counters into groups. Alternatively, the activity could be completed on individual whiteboards with students using whiteboard markers to circle the groups. Encourage students to record the equations. Repeat with a number of different multiplication equations.

DIVISION ON NUMBER LINES

- For students who require a challenge, have them use number lines to draw and complete division equations based on tables facts. Have students represent both divisions on the same number line. For example: for 6 × 4, students need to represent 24 ÷ 6 and then 24 ÷ 4 on the same number line.

REFLECTION

Select from the following to suit your class and their learning outcomes:

- Have students share their array explorations from Independent Tasks, Task 1, with the group. Look for interesting representations, and students who have looked at all alternatives.
- Have students share their electronic representation of arrays from Independent Tasks, Task 2. Look for interesting representations.
- Have students look at the array in a prop such as a muffin tray. Have them identify all of the multiplication facts related to the array and then all of the related division facts. Have students record their responses on the whiteboard.

Home Tasks

Select from the possible Home Tasks:

- Have students take home their multiplication grids to help when practising their tables.
- Have students hunt for arrays around the house, e.g. packaging, cartons, storage materials. Have students record what they found and bring it to class. Incorporate these ideas into the lessons.

Assessment

- Have students complete **Student Assessment p. 47**.
- Review with students **Assessment Task Card 4.11**.

During the three lessons:

- Make a copy or note of the work done in Lesson Plan 3, Independent Tasks, Task 1, showing students' specific abilities.
- Make note of the tables sets students are struggling with, and those they have mastered.
- Make note of the depth of the fact families students are producing in class activities and on Student Book pages.

Recommendations for Future Learning

Specific to Student Assessment p. 47; if the student is experiencing some difficulty:

Q 1 Review what an array is and how it is represented and drawn. Review the role of the columns and rows, and how this links to the written equation. This could also be reviewed with hands-on materials, such as counters.

Q 2–4 Revisit tables and answers using materials such as BLM 3 'Tables Chart 1', BLM 4 'Tables Chart 2' and NTO 4.1 'Hundred Chart'. Look at the links between multiplication and division and how this relates to the array.

Q 5 Review how to circle the correct groups to indicate the divisions. Discuss the two different representations of division, and how this relates to the array.

If the student has not achieved the recommended skills for this unit:

1. See **Assessment Task Card 4.11** for specific recommendations.
2. Have the student use BLM 3 'Tables Chart 1' and BLM 4 'Tables Chart 2' to practise tables. They can also work with their own multiplication grids.
3. Use the tables charts and multiplication grids to help in understanding the link between multiplication and division.
4. Use concrete materials, such as counters or egg cartons, to give a 'real feel' to arrays.
5. Review *Nelson Maths: Australian Curriculum NSW Year 3* Units 22 and 24.

If the student has achieved the recommended skills and these skills are firmly established, consider:

1. Having the student complete 'Multiplication Facts to Solve Division' from *Nelson Maths Building Mental Strategies Big Book 4*, pp. 60–61.
2. Moving forward to *Nelson Maths: Australian Curriculum NSW Year 5* Units 10 or 11.
3. Extending the student in any of the listed activities by moving to the 11 and 12 times tables or to tables multiplied by a factor of 10 or 100.

Unit 12 Mapping

Measurement and Geometry

Position MA2-17MG uses simple maps and grids to represent position and follow routes, including using compass directions

direction, features, legend, map, scale

LESSON PLAN 1

TUNING IN

WHAT CAN YOU SEE?

You will need: an IWB or data projector, an interactive map tool

Display a map on the IWB or data projector – it could be of your local area or an area of interest. Ask, 'What is this a map of? What can you see on the map?' Draw out aspects such as towns, roads, parks and schools. Ask, 'How do you know this is a road? What is the difference between the different-coloured roads?'

WHOLE-CLASS INTRODUCTION

A LEGEND

You will need: an IWB or data projector, an interactive map tool

Following the observations made in the Tuning In activity, have students draw and label each of the aspects they observed on the board. For example: ▬▬▬ = main road. Introduce the term 'legend'. Then examine a different map to see if the legend works. Have students adjust and adapt accordingly.

INDEPENDENT TASKS

Note: Choose from Tasks 1, 2 or 3.

You will need: photocopies of a relevant map, coloured pencils or felt pens, LO: *L8304 'Pirate treasure hunt: eight challenges'*, Student Book p. 48 'At the Fun Park'

TASK 1: CREATE MY OWN LEGEND

Give students a photocopy of a selected map. This may be of the local area, related to a theme of study (e.g. the Ballarat goldfields) or linked to an upcoming excursion. Have students create their own legend for the map, using colour and the features on the map.

TASK 2: INTERACTIVE TASK

Have students work independently on computers, using LO: *L8304 'Pirate treasure hunt: eight challenges'* to use a map and different strategies to solve the clues and find the treasure.

TASK 3: STUDENT BOOK p. 48 *'At the Fun Park'*

TEACHING GROUP

You will need: copies of maps

USING MAPS

- For students who require support, give them copies of maps and spend time as a group identifying the different features. Discuss the need for the legend to represent the information in the map. Scaffold students by drawing the symbols for the features and then having them write in the descriptors.

CREATING MAPS

- For students who require a challenge, have them draw their own map of the school from memory, including a legend.

REFLECTION

Select from the following to suit your class and their learning outcomes:

- Have students share the legends they created in Independent Tasks, Task 1. Compare different students' responses. Make sure students have included all of the main features on the map, e.g. roads, parks.

- Have students from the Teaching Group 'Creating Maps' share their maps of the school. Ask, 'What was difficult about the task? How would we know if the map was accurate?'
- Have students create a map showing how they travel from home to school, including a legend.

LESSON PLAN **2**

TUNING IN

LOOKING AT SCALE

You will need: an IWB or data projector, an interactive map tool

Display a map of Australia on the IWB or data projector and have students examine the scale as you zoom in and out of a location. Begin with the whole of Australia and look at the scale. Ask, 'What does it mean?' Have students come to the board and measure with a ruler to see that Australia is more than 4000 km wide. Zoom in to your local area – either the city or district, again examining the scale, and then to your suburb or town to the school location. Repeat with a different country such as New Zealand or China.

WHOLE-CLASS INTRODUCTION

COMPARING SCALE

You will need: an IWB or data projector, an interactive map tool

Review or introduce how to read scale, e.g. 1:100 cm, and the link to the visual representation. Look at another example, e.g. 1:1 km, and how this is different. Note: keep the scales simple, i.e. factors of 10, 100 or 1000 rather than 2.5:500 m. Link this back to the examples on the interactive map tool.

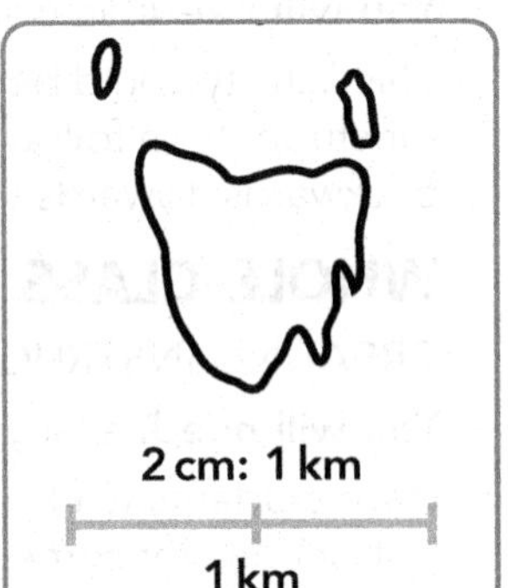

INDEPENDENT TASKS

Note: Choose from Tasks 1, 2 or 3.

You will need: copies of two maps for comparison purposes (e.g. a city and a rural area or an Australian city and an overseas city), poster paper, rulers, tape measures, scrap paper, Student Book p. 49 'The Science Centre'

TASK 1: MAKING COMPARISONS

Give pairs of students copies of two maps to compare. Have students stick their maps onto poster paper. Students identify the scale and write it under each map, as well as the major features of each map. Then they write a description about the differences and similarities between the maps. This activity could be varied by having the pairs examine different sets of maps or it could be linked to students' LOTE study.

TASK 2: INTERACTIVE TASK

Have students work in pairs on computers to create a scaled map of the classroom. Give students tools, such as rulers, tape measures and scrap paper. This task can be varied by providing a small or a large amount of instruction. Allow students to select the appropriate tool or software to complete the task.

TASK 3: STUDENT BOOK p. 49 *'The Science Centre'*

TEACHING GROUP

You will need: assorted hands-on equipment (e.g. Unifix blocks, counters), a digital camera, http://www.spatialgenie.edu.au/spatialgenie/, *M009137 'Measuring the Distance on a Map – Teacher Resource'* (this can be found on the econtent site: http://econtent.thelearningfederation.edu.au/ec/p/home)

CREATING A CLASSROOM

- For students who require support, give them a selection of hands-on equipment such as Unifix blocks and counters. This activity could be completed individually or in pairs. Have them create a model of the classroom, using appropriately sized objects to represent different aspects, e.g. two Unifix blocks joined together might represent a table, and one block would represent a chair. Take a photo of the completed model. Then have students create a relevant legend for their classroom model, either on hard copy or electronically.

SPATIAL GENIE

- For students who require a challenge, have them work on computers to use the mapping tool Spatial Genie (http://www.spatialgenie.edu.au/spatialgenie/). Have students explore scale on the maps. This can be supported with *M009137 'Measuring the Distance on a Map – Teacher Resource'*.

REFLECTION

Select from the following to suit your class and their learning outcomes:

- Have pairs of students share their map comparisons from Independent Tasks, Task 1. Ask questions about what students discovered in terms of similarities and differences.
- Invite students to share the classroom maps they made in Independent Tasks, Task 2. Ask, 'What was tricky to work out? If you had to do the task again, how would you approach it? Would you start at the same place or somewhere different?' Students could also be invited to share the models they made in the Teaching Group 'Creating a Classroom'.
- Revisit some maps, perhaps from an atlas or a street directory, and examine the scale of different maps and presentations. Ask, 'What is the scale of this map? What are the main features of this map?'

LESSON PLAN 3

TUNING IN

FOLLOWING DIRECTIONS

You will need: an open space

This activity could be done outside. Ask each student to stand in a space of their own. Then give directions for students to follow. For example: 'Take two steps forwards. Take three steps to the right. Take one step backwards, towards the door.' Ask students to comment on what they see happening.

WHOLE-CLASS INTRODUCTION

CREATING INSTRUCTIONS

You will need: a large outdoor space such as a basketball court

Have students work in pairs in a large outdoor space. Tell students that they need to create a list of instructions for someone to move from one location, e.g. the basketball ring, to another location, e.g. the centre circle. Impose some restrictions, e.g. no more than five steps forwards can be taken; include some backwards directions. Remind students to include the starting location.

INDEPENDENT TASKS

Note: Choose from Tasks 1, 2 or 3.

You will need: a printed map (e.g. the local area from the internet), coloured pencils or felt pens, LO: *L3535 'Ladybird mazes'*, Student Book p. 50 'Things to Do'

TASK 1: HERE AND THERE

Give each student a map. Have students write a statement of how to get from one location, e.g. school, to another location, e.g. the local shops. Have them write an alternative route as well. Then allow students to select another two locations and write the directions. Have students mark the different starting points and destinations in different colours on the map.

TASK 2: INTERACTIVE TASK

Have students work individually on computers, using LO: *L3535 'Ladybird mazes'* to manoeuvre a ladybird through a maze using forwards and backwards arrows and rotations of 90 and 45 degrees.

TASK 3: STUDENT BOOK p. 50 *'Things to Do'*

TEACHING GROUP

You will need: photocopies of a map, internet or library access

CAN YOU WORK IT OUT?

- For students who require support, have them work in pairs and give them each a copy of the same map. One student circles two locations on their map then, without showing their map, gives their partner the starting location and directions to follow to finish at the same destination. The partner draws the route on their own map. The aim is for students to work on their descriptive skills, e.g. 'Follow the road towards town'. Have students swap roles and repeat.

MAPS OF THE WORLD AND BEYOND

- For students who require a challenge, have them use the internet or the library to research different types of maps. These could be historic maps, maps in different languages or maps of space or the moon. Have students collect information as they research. Have them consider aspects such as scales and legends.

REFLECTION

Select from the following to suit your class and their learning outcomes:

- Have students swap the instructions they made in the Whole-Class Introduction and try to follow them. Ask, 'Where did you start? Where did you finish? What was helpful about the instructions?' Check answers.
- Have students reflect on what is important when providing directions. Display these ideas as a reminder.
- Have the class examine animated maps. Examples can be found at the Sovereign Hill, Australia Zoo and Melbourne Zoo websites. Ask, 'What is different about these maps?' Have students identify scale and legends.
- Invite students from the Teaching Group 'Maps of the World and Beyond' to share what they found when researching maps. Examine what is different and what is the same about various maps.

Home Tasks

Select from the possible Home Tasks:

- On a sheet of paper, have students create a pirate map. Encourage them to include features such as a scale and a legend. To extend the activity, students could create some directions to find the treasure!
- Have students create a scale drawing of their bedroom; they may include a legend.

Assessment

- Have students complete **Student Assessment p. 51**.
- Review with students **Assessment Task Card 4.12**.

During the three lessons:

- Collect students' maps with their own determined legends from Lesson Plan 1, Independent Tasks, Task 1, as a sample for their portfolios.
- Make note of students who completed assisted or extension tasks and what they achieved or still struggled with during that time.
- Collect digital materials such as those created in Lesson Plan 2, Independent Tasks, Task 2, and add to students' digital portfolios as an example of work on scale and mapping.

Recommendations for Future Learning

Specific to Student Assessment p. 51; if the student is experiencing some difficulty:

Q 1, 3 & 4 Review the ideas behind giving directions. Collect a set of key terms to aid the student, e.g. 'next to', 'to the left', 'to the right', and display in the room or add to the student's maths toolkit. Begin with simple instructions and move to more complex ones. Remind the student to re-read instructions that they write to ensure they make sense.

Q 2 Revisit the idea of a legend – what it looks like and what information it shows. Look at a number of examples with similar elements, e.g. roads, toilets, parks. You could create a chart of common symbols to help the student.

Q 5 Revisit scale, and review how different maps have different scales, dependent on factors such as the size of the location, and how close the view of the map is. Work with real-life models and then move to paper/one-dimensional representations.

If the student has not achieved the recommended skills for this unit:

1. See **Assessment Task Card 4.12** for specific recommendations.
2. Continue looking at different types of maps, e.g. brochures, digital representations.
3. Take a digital photo of an area or space, such as the playground. Display the photo and show the student how to develop a map from this.
4. Collect a set of common symbols used in map legends, to help the student.
5. Give the student simple instructions to follow, highlighting key terms.
6. Review *Nelson Maths: Australian Curriculum NSW Year 3* Unit 7.

If the student has achieved the recommended skills and these skills are firmly established, consider:

1. Having the student work with more complex and unfamiliar maps.
2. Moving forward to *Nelson Maths: Australian Curriculum NSW Year 5* Unit 12.
3. Having the student aim for more accuracy when using/creating scale and creating legends.
4. Having the student begin to use simple grid referencing.

Unit 13 Addition and Subtraction

Number and Algebra

Addition and subtraction MA2-5NA uses mental and written strategies for addition and subtraction involving two-, three-, four- and five-digit numbers

addition, doubles, equations, near-doubles, number lines, place value, subtraction

LESSON PLAN 1

TUNING IN

ADDING

You will need: up to four large dice

Have students sit in a circle. Roll two, three or four dice (depending on students' abilities) and move around the circle, having students complete the addition of the numbers rolled. Stop occasionally to ask students the strategies they are using to complete the addition.

WHOLE-CLASS INTRODUCTION

ADDITION STRATEGIES

Write addition equations on the board to highlight different strategies. For example: 24 + 66 = to illustrate adding to tens; 18 + 17 = to illustrate doubles or near-doubles; 42 + 15 = to illustrate adding tens and ones together. Initially, work with numbers less than 100 to consolidate ideas and strategies. Name the strategies used to help students to be able to articulate their work and learning.

INDEPENDENT TASKS

Note: Choose from Tasks 1, 2 or 3.

You will need: dice, LO: *L871 'Wishball challenge: whole numbers'*, Student Book p. 52 'Addition'

TASK 1: 99 OR BUST

Give pairs of students a dice. Have students take turns to roll the dice and keep a running total of the numbers they each roll. The winner is the first student to get the closest to 100 without going over. At any time the student can decide to 'sit' but can't re-enter the game. If the student goes over 100 they immediately lose. Play a number of rounds. To vary the activity, students can play with two dice, enabling them to create 2-digit numbers, and can play for a target of 199.

TASK 2: INTERACTIVE TASK

Have students work independently on computers, using LO: *L871 'Wishball challenge: whole numbers'* to add or subtract numbers based on place-value position.

TASK 3: STUDENT BOOK p. 52 *'Addition'*

TEACHING GROUP

You will need: dice, concrete materials (e.g. MAB, counters), calculators

PRACTISING ADDITION

- For students who require support, organise them into pairs and give each pair four dice (two each). Have each student roll their dice and make a 2-digit number. Then pairs add their two numbers together. Provide concrete materials to help if needed. Repeat a number of times, with students working together and practising strategies. When complete, allow students to check answers with calculators.

MAKING NUMBERS

- For students who require a challenge, have them use the digits 1 to 9 and any operation to make all of the numbers between 40 and 80. Challenge students to use the fewest numbers and operations each time. To extend the activity, move the numbers into the hundreds.

REFLECTION

Select from the following to suit your class and their learning outcomes:

- Have students write a written reflection of all of the terms they understand to mean addition, and their favourite addition strategy to use and why.
- Have students from the Teaching Group 'Making Numbers' share their findings. Ask, 'Did you make all the numbers? What was the most challenging number to make and why?'
- Ask two students to each provide a 3-digit number. Write them on the board and have a class discussion about how to solve the addition of these two numbers. Ask, 'What strategies could be used?'

LESSON PLAN 2

TUNING IN

99 OR BUST REVERSED

You will need: a 10-sided dice

Play '99 or Bust' in reverse and with a 10-sided dice. (See Lesson Plan 1, Independent Tasks, Task 1.) Start at 99 and subtract each of the rolled numbers with the target number being 1. Play a number of rounds. Ask, 'Which version of the game is easier to play? Why?'

WHOLE-CLASS INTRODUCTION

FACT FAMILES

Write subtraction equations on the board to discuss the different strategies. For example: 51 – 29 = to illustrate the compensation strategy; 68 – 26 = to illustrate the split strategy. Also present a number of examples requiring trading, and discuss. Talk about horizontal and vertical representations.

INDEPENDENT TASKS

Note: Choose from Tasks 1, 2 or 3.

You will need: 10-sided dice, poster paper, LO: *L99 'The take-away bar: generate easy subtractions'*, Student Book p. 53 'Subtraction'

TASK 1: COMPLETING PATTERNS

Have students investigate subtraction patterns based on multiples of 10. For example:

4 – 2 = 2

40 – 20 = 20

400 – 200 = 200

4000 – 2000 = 2000

Students could be given the starting equation or they could generate their own, using 10-sided dice. Have students present their findings on a poster, describing the patterns they created.

TASK 2: INTERACTIVE TASK 

Have students work independently on computers, using LO: *L99 'The take-away bar: generate easy subtractions'* to use a slider to help their understanding of subtraction equations. Note: there are various versions of this Learning Object to cater for different abilities.

TASK 3: STUDENT BOOK p. 53 ***'Subtraction'***

TEACHING GROUP

You will need: calculators, pieces of card

SUBTRACTION PATTERNS

- For students who require support, work on subtraction patterns such as those shown at right. Allow students to use calculators, then have them try to work out the subtractions based on the patterns. Repeat a number of times, altering the level of equations depending on students' abilities.

–	56	66	76	86
20				

FINDING THE DIFFERENCE

- For students who require a challenge, pose the problem: 'Look at these numbers: 2, 8, 52, 91. Which two of these numbers have a difference of 83?' Have students work out the answer. Then have students write five versions of this problem on pieces of card, using 3- or 4-digit numbers. Have students write the answer on the back of each card. When complete, have students swap their cards and solve each other's problems.

REFLECTION

Select from the following to suit your class and their learning outcomes:

- Invite students to share their subtraction patterns from Independent Tasks, Task 1. Have students explain the pattern they found in their particular series of numbers.
- Have students reflect on the subtractions from Independent Tasks, Task 2. Ask, 'What did you find difficult about the activity? What was easy?'
- Have some students from the Teaching Group 'Finding the Difference' read out the problems they created for the rest of the class to try to solve. Alternatively, a number of them could be copied and provided as a worksheet to the rest of the group.

LESSON PLAN 3

TUNING IN

BINGO DIVISION

You will need: BLM 6 'Numbers and Words 1', BLM 7 'Numbers and Words 2', BLM 8 'Numbers and Words 3', felt pens

Copy and cut up the number cards from BLM 6 'Numbers and Words 1, BLM 7 'Numbers and Words 2' and BLM 8 'Numbers and Words 3'. Give one card to each student. Have students use the cards to line up in order from smallest to largest. Then select a student and take an amount off their number, e.g. 50. Have them work out the new value, write it on the back of the card and reposition themselves in the line. Repeat a number of times. Ask students what is happening.

WHOLE-CLASS INTRODUCTION

LARGE NUMBER LINE: ADDITION AND SUBTRACTION

You will need: different-coloured chalk

Take students outside. Use a chalk line on the ground or on a wall, or a line on the basketball court, to create an open number line for students. Nominate whether students are to create an addition or a subtraction equation. Give one student a starting number and have them stand on the line. Then give a number to another student and have the group work together to complete the number line to solve the equation. Have students record their working in one colour of chalk. Repeat a number of times, looking at both addition and subtraction equations, using a different chalk colour each time. Remind students how to use number lines, e.g. drawing arrows, recording the jump values, recording the equation and answer.

INDEPENDENT TASKS

Note: Choose from Tasks 1, 2 or 3.

You will need: coloured chalk, a digital camera or flip camera, Student Book p. 54 'Addition and Subtraction Number Lines'

TASK 1: ADDITION AND SUBTRACTION NUMBER LINES

Give pairs of students addition and subtraction equations, according to their abilities, to solve using open number lines. Have students draw the number lines with chalk on the ground or external walls. When complete take digital photos of the equations and number lines. This activity could be varied using a flip camera, and having students record a description of how they solved the equation using the number line.

TASK 2: INTERACTIVE TASK

Have students use computers to create two posters, one on addition and one on subtraction. Students should include other terms that mean addition and subtraction, strategies they are familiar with for each operation and examples of the strategies. Encourage students to be creative.

TASK 3: STUDENT BOOK p. 54 *'Addition and Subtraction Number Lines'*

TEACHING GROUP

You will need: NTO 4.4 'Number Line: Patterns', 10-sided dice

ELECTRONIC NUMBER LINE

- For students who require support, have them use NTO 4.4 'Number Line: Patterns' to practise addition and subtraction equations. Have students record the equations in their books. After a number of examples, have students revisit their recorded equations and draw the relevant open number lines. Remind students how to use number lines, e.g. drawing arrows, placing the largest number first, labelling the jumps, recording the equation and answer.

ROLLING DICE

- For students who require a challenge, give pairs of students two 10-sided dice. Have each student list down a page the numbers 0 to 20. Students take turns to roll the dice and then use the numbers to make an equation between 0 and 20, using either addition or subtraction, e.g. 5 + 6 = 11. Have the student record the equation next to the number on their list. The aim of the game is to make each number on the list. If a student cannot make a number they need on the roll, the game progresses and they do not receive an extra turn.

REFLECTION

Select from the following to suit your class and their learning outcomes:

- As a class, share the photos of the chalk number lines made in Independent Tasks, Task 1. Look at the number lines that are clear and invite students to comment on their own work.
- Have students share the posters they made in Independent Tasks, Task 2. After sharing, students may wish to add ideas to their own posters.
- Create a class poster of all the different addition and subtraction strategies that students use. Provide examples with each strategy and display the poster as a reference.

Home Tasks

Select from the possible Home Tasks:

- Have students look at advertisements in newspapers or on television. Students note three items to buy and find the total cost. Have students bring their work to school to share.
- Have students teach their parents or carers about open number lines, showing an example of an addition and a subtraction equation. Invite students to share their experiences back in class.

Assessment

- Have students complete **Student Assessment p. 55**.
- Review with students **Assessment Task Card 4.13**.

During the three lessons:

- Make a copy of students' written reflections on addition from Lesson Plan 1.
- Collect copies of digital materials, such as the digital posters or photos of number lines from Lesson Plan 3, Independent Tasks, to add to students' digital portfolios.
- Make note of students who are still struggling with basic concepts and spend time reviewing basic facts.

Recommendations for Future Learning

Specific to Student Assessment p. 55; if the student is experiencing some difficulty:

Q 1 Review addition strategies with the student. Use modelling equipment, e.g. MAB, to illustrate the addition strategies.

Q 2–3 Use modelling equipment, e.g. MAB and bundling sticks, to demonstrate subtraction strategies. Spend time working with smaller numbers, e.g. 2-digit numbers, before moving to larger numbers. Spend time consolidating subtraction ideas before moving to subtraction with trading.

Q 4 Review patterns using basic addition and subtraction equations. Review Independent Tasks, Task 2, from Lesson Plan 2.

Q 5 Use NTO 4.4 'Number Line: Patterns' to revisit concepts about using number lines. Highlight aspects such as the placement of the first number; breaking the second number into place-value components, e.g. hundreds, tens and ones; and labelling the diagram.

If the student has not achieved the recommended skills for this unit:

1. See **Assessment Task Card 4.13** for specific recommendations.
2. Review basic number facts to 20, consolidating basic ideas, e.g. doubles, making to 10.
3. Work with numbers to 100 and then 1000. Start with addition and subtraction equations that do not require trading.
4. Use concrete materials, e.g. bundling sticks and MAB, to illustrate trading concepts.
5. Review *Nelson Maths: Australian Curriculum NSW Year 3* Unit 14.

If the student has achieved the recommended skills and these skills are firmly established, consider:

1. Having the student complete 'Adding Place-Value Units' and 'Using Open Number Lines for Addition', and 'Counting Back' and 'Counting Up To' from *Nelson Maths Building Mental Strategies Big Book 4*, pp. 26–29, 40–43.
2. Moving forward to *Nelson Maths: Australian Curriculum NSW Year 5* Unit 2.

Unit 14 Perimeter and Area

Measurement and Geometry

Length MA2-9MG measures, records, compares and estimates lengths, distances and perimeters in metres, centimetres and millimetres, and measures, compares and records temperatures

Area MA2-10MG measures, records, compares and estimates areas using square centimetres and square metres

ML area, centimetres, irregular shapes, length, measure, metres, perimeter, regular shapes, units, width

LESSON PLAN 1

TUNING IN

PERIMETER OF A BOOK

You will need: rulers

Ask students to measure the perimeter of a book with a ruler. Have students record each of the side lengths and draw a simple diagram. Review terms such as 'length' and 'width' and the units. Check to ensure students use rulers and scales correctly. Review the term 'perimeter' and how to find it.

WHOLE-CLASS INTRODUCTION

IRREGULAR SHAPES

You will need: rulers

Draw an irregular shape on the board, such as the one at right.
Ask, 'How would we find the perimeter of this shape?' Depending on students' abilities, the diagram could be labelled with lengths, or students could come to the board and measure with a ruler. Have students find the perimeter. Ensure they label each of the side lengths, and that they show the addition equation to find the perimeter. This activity could be extended by providing a diagram with missing values.

INDEPENDENT TASKS

Note: Choose from Tasks 1, 2 or 3.

You will need: flat blocks (e.g. pattern blocks), A3 paper, rulers, LO: *L3528 'Geoboard'*, geoboards, Student Book p. 56 'Perimeter of Irregular Shapes', BLM 26 'Square Dot Paper'

TASK 1: CREATING IRREGULAR SHAPES

Give students some flat blocks, e.g. pattern blocks. Have them create five different irregular shapes with the blocks on a sheet of A3 paper. Have students trace around the blocks carefully, and then find the perimeter of each of their irregular shapes. Ensure students label each of the side lengths correctly and with the correct units. When complete, students may wish to colour their shapes, and add a heading to their work. This activity could be extended by setting requirements for students, e.g. make a five-sided irregular shape.

TASK 2: INTERACTIVE TASK

Have students work independently on computers, using LO: *L3528 'Geoboard'* to complete a number of tasks. This could be complemented in the classroom with a set of 'real' geoboards and having students look at the perimeter of both regular and irregular shapes.

TASK 3: STUDENT BOOK p. 56 *'Perimeter of Irregular Shapes'*

Note: students will require BLM 26 'Square Dot Paper' for this activity.

TEACHING GROUP

You will need: flat blocks (e.g. pattern blocks), rulers, calculators, BLM 26 'Square Dot Paper'

PRACTISING PERIMETER

- For students who require support, give them a number of flat blocks, e.g. pattern blocks. Have them place the blocks on a sheet of paper, and trace carefully around each one. Then have students measure each of the side lengths with a ruler, and label the shape. Repeat a number of times with different shapes, e.g. a triangle, a square and a rectangle. Then have students work out the perimeter of each shape. Students could use calculators to help with the adding process.

WORKING BACKWARDS

- For students who require a challenge, give them a perimeter value, e.g. 30 cm, and have them investigate how many shapes (regular and irregular) they can construct with the perimeter value. Students could use BLM 26 'Square Dot Paper' to help with the investigation. Students could be extended further by giving them a larger value, e.g. 100 cm.

REFLECTION

Select from the following to suit your class and their learning outcomes:

- Give the class a number of sheets of newspaper and have them create an irregular shape on the floor. As a group, have them find each of the side lengths of the shape, and then the perimeter. Metre rulers could be used. Take a photo of the large irregular shape.
- Have students share the shapes they made in Independent Tasks, Task 1, and the perimeters of the shapes. Ask, 'How did you find that side length? How did you know those two side lengths were the same? How could you check your answer?'
- Have students from the Teaching Group 'Working Backwards' share the shapes they made. Have them explain how they went about solving the problem.

LESSON PLAN 2

TUNING IN

AREA OF CLASSROOM FLOOR

You will need: sheets of newspaper pre-cut into areas of 1 m^2

Have students work as a group to find how many sheets of newspaper cover the classroom floor. Tell students that each sheet measures an area of 1 m^2 and then have them work out the approximate area of the classroom floor.

WHOLE-CLASS INTRODUCTION

AREA IN SQUARES

You will need: grid paper on the IWB (this is readily available on IWB software)

Show students two shapes drawn over the grid paper on the IWB, such as those at right. Discuss how to find area by counting squares, and then how to write the answer. Have students compare the areas of the two shapes. Ask, 'Which is larger?' Investigate a number of examples. Note: this could be completed on a whiteboard, with hand-drawn diagrams.

INDEPENDENT TASKS

Note: Choose from Tasks 1, 2 or 3.

You will need: BLM 17 '1 cm Grid Paper', LO: *L139 'Area counting with Coco'*, Student Book p. 57 'Island Areas'

TASK 1: THE AREA OF MY HAND

Give students a copy of BLM 17 '1 cm Grid Paper'. Have students trace around their hand on the grid, then count squares to find the area of their hand. About five minutes into the activity, ask students to stop work and share different techniques for completing the task, e.g. numbering squares as you count, colouring squares as you count, creating and counting rectangular areas first. Students then finish finding the area.

TASK 2: INTERACTIVE TASK

Have students work independently on computers, using LO: *L139 'Area counting with Coco'* to find the area of regular and compound shapes. Note: students will be using the formula to find area in this activity.

TASK 3: STUDENT BOOK p. 57 *'Island Areas'*

TEACHING GROUP

You will need: flat blocks (e.g. pattern blocks), BLM 17 '1 cm Grid Paper'

AREAS OF BLOCKS

- For students who require support, give them a number of flat blocks, e.g. pattern blocks. Have students place the blocks on a copy of BLM 17 '1 cm Grid Paper' and trace around the blocks. Ensure students line up one of the straight edges of the block with the straight lines on the grid paper. Then students count squares to find the area of each shape. Have them label the centre of the block with the area, using the correct units.

DETERMINING THE AREA

- For students who require a challenge, give them a copy of BLM 17 '1 cm Grid Paper' and an area value, e.g. 76 cm^2. Have students investigate how many regular and compound shapes they can make with that area value. Students can work independently and then in pairs. Students could be further extended by giving them a larger value, e.g. 200 cm^2.

REFLECTION

Select from the following to suit your class and their learning outcomes:

- Have students compare the hand areas they calculated in Independent Tasks, Task 1. Who has the smallest hand area? The largest? Whose hand areas are the same?
- Review Student Book p. 57 'Island Areas'. Share strategies students used to complete the activity, and check answers.
- Have students write a reflection on how to find the area of a shape. Encourage them to include diagrams in their reflection.

LESSON PLAN 3

TUNING IN

MAKING 1 CM^2 AND 1 M^2

You will need: newspaper, scissors, tape

Divide students into groups and have each student make pieces of newspaper with areas of 1 cm^2 and 1 m^2. Have the groups share their results. Compare the two areas. Ask, 'What do you notice? What is the same? What is different?' Draw out the importance of units. Retain the 1 cm^2 and 1 m^2 shapes.

WHOLE-CLASS INTRODUCTION

AREA OF THE BASKETBALL COURT

You will need: the 1 m^2 areas from Tuning In

As a class, have students find the area of the basketball court using their 1 m^2 shapes from the Tuning In activity. Have students compare this to the area of their classroom (see Lesson Plan 2, Tuning In). Ask, 'Which area is larger? How are the units important?'

INDEPENDENT TASKS

Note: Choose from Tasks 1, 2 or 3.

You will need: rulers, LO: *L383 'Finding the area of compound shapes'*, Student Book p. 58 'Finding Area', BLM 17 '1 cm Grid Paper'

TASK 1: AREA AROUND THE CLASSROOM

Have students draw up a table as at right. Have students move around the room looking for objects, measuring them and entering them into the table. Their challenge is to find two objects for each column of the table.

Less Than 1 cm^2		
Between 1 cm^2 and 1 m^2		
Greater Than 1 m^2		

TASK 2: INTERACTIVE TASK

Have students work independently on computers, using LO: *L383 'Finding the area of compound shapes'* to find the area of compound shapes, by breaking them into regular shapes. Note: students will be using the formula to find area in this activity.

TASK 3: STUDENT BOOK p. 58 *'Finding Area'*

Note: students will require BLM 17 '1 cm Grid Paper' for this activity.

TEACHING GROUP

You will need: BLM 17 '1 cm Grid Paper', BLM 27 'Island Challenge', interlocking cubes, rulers

COMPOUND SHAPES AND AREA

- For students who require support, give them a copy of BLM 17 '1 cm Grid Paper' with some compound shapes drawn on it. Have students divide the compound shapes into familiar squares and rectangles. Students count squares to find the area of each shape and then add the areas together to find the total area. Students could use interlocking cubes to create the compound shapes and then physically pull them apart to make the more familiar shapes of squares and rectangles to aid understanding. If students are confident with the process, move on to some simple compound shapes with no grids.

ISLAND CHALLENGE

- For students who require a challenge, have them complete BLM 27 'Island Challenge'.

REFLECTION

Select from the following to suit your class and their learning outcomes:

- Have students share the tables they created in Independent Tasks, Task 1. Ask students which were the most challenging areas to find. Share unusual responses.
- As a class, go through the answers to Student Book p. 58 'Finding Area'.
- Draw a large triangle on the board and have students brainstorm how they might find the area of this shape.

Home Tasks

Select from the possible Home Tasks:

- Have students look for three irregular shapes, record the names of the items and their perimeters.
- Have students take home their 1 m^2 shapes and measure the area of their bedroom floor. This activity could be extended by having students draw a diagram and labelling both the perimeter and area.

Assessment

- Have students complete **Student Assessment p. 59**.
- Review with students **Assessment Task Card 4.14**.

During the three lessons:

- Collect students' written reflections from Lesson Plan 2 on finding areas of shapes to add to their portfolios.
- Take photos of students completing activities to add to their digital portfolios. Students could be invited to write a comment about the activity and their learning.
- Note students' involvement in and ability to complete the interactive activities.

Recommendations for Future Learning

Specific to Student Assessment p. 59; if the student is experiencing some difficulty:

Q 1 Revisit the finding of perimeter and area by tracing shapes onto grid paper, and then measuring with a ruler or counting squares.

Q 2–4 Revisit the difference in units for area. Have the student compare their 1 cm^2 and 1 m^2 pieces of paper.

If the student has not achieved the recommended skills for this unit:

1. See **Assessment Task Card 4.14** for specific recommendations.
2. Have the student work with simple shapes, e.g. squares and rectangles, before moving to compound shapes.
3. Have the student continue working with the 1 cm grid paper to find areas. Continue the process of tracing the shape and then counting squares.
4. Review *Nelson Maths: Australian Curriculum NSW Year 3* Unit 4.

If the student has achieved the recommended skills and these skills are firmly established, consider:

1. Having the student work with more complex shapes, including triangles, trapeziums and parallelograms.
2. Moving forward to *Nelson Maths: Australian Curriculum NSW Year 5* Units 14 or 15.
3. Challenging the student by looking at informal methods of finding the perimeter and area of a circle.

Multiplication and Division Strategies

Number and Algebra

Multiplication and division MA2-6NA uses mental and informal written strategies for multiplication and division

division, doubling, halving, multiplication, place value, strategy

LESSON PLAN 1

TUNING IN

MATCHING EQUATIONS

You will need: BLM 25 'Matching Equations'

Photocopy and cut up the cards from BLM 25 'Matching Equations'. Give students one card each. Have them find their partner by finding the reversed equivalent equation, e.g. $2 \times 6 = 12$ and $12 = 6 \times 2$. Ask, 'What do you notice?' Repeat a number of times, mixing up the cards each time.

WHOLE-CLASS INTRODUCTION

COMMUTATIVITY

You will need: BLM 3 'Tables Chart 1', BLM 4 'Tables Chart 2'

Write a number of multiplication equations on the board, e.g. $6 \times 5 = 30$, $7 \times 8 = 56$, and give students a copy of BLM 3 'Tables Chart 1' and BLM 4 'Tables Chart 2'. Have them find all of the reversed equivalent equations and colour them in different colours. Ask, 'How many can you find?'

INDEPENDENT TASKS

Note: Choose from Tasks 1, 2 or 3.

You will need: poster paper, LO: *L61 'The multiplier: make your own easy multiplications'*, Student Book p. 60 'Commutative Arrays'

TASK 1: COMMUTATIVE ARRAYS

Give pairs of students a sheet of poster paper. Have students select pairs of reversed equivalent multiplication tables, list the two equations and draw an array to represent each. Have students complete a minimum of five pairs.

TASK 2: INTERACTIVE TASK

Have students work independently on computers, using a partitioning tool on LO: *L61 'The multiplier: make your own easy multiplications'* to help solve their own multiplications.

TASK 3: STUDENT BOOK p. 60 *'Commutative Arrays'*

TEACHING GROUP

You will need: LO: *L106 'The array'*, calculators

PRACTISING MULTIPLICATION

- For students who require support, have them work independently on computers, using an array-building tool on LO: *L106 'The array'* to help solve multiplications. Students will examine relationships between rows, columns and areas in arrays.

LARGER NUMBERS

- For students who require a challenge, have them explore the concept of commutativity with larger numbers and multiplications, e.g. $20 \times 50 =$ and $50 \times 20 =$. Students may need to use calculators. Have students record their ideas and findings. They can also explore the breaking down of numbers, e.g. $50 \times 20 = 10 \times 5 \times 2 \times 10$, and if this also applies.

REFLECTION

Select from the following to suit your class and their learning outcomes:

- Have students share the posters they made in Independent Tasks, Task 1. Look for students who have produced similar ideas and representations. Ask questions to clarify students' understanding of the construction of the arrays.
- Have students from the Teaching Group 'Larger Numbers' share their investigation of the multiplication of larger numbers. Ask, 'Did it still work?' Have them present their exploration of the breaking down of numbers in multiplication equations.
- Give students a number of multiplication equations on the board, e.g. 12 × 10 = 120 (at least 10, dependent on students' abilities). Ask the reversed equation, e.g. 10 × 12 =, and have students identify the answers.

LESSON PLAN 2

TUNING IN

DOUBLING AND HALVING GROUPS

Have students play 'Clumps', where a number is called and they get into groups of that size. Move to calling only even numbers, then have students halve or double the groups. Play a number of rounds.

WHOLE-CLASS INTRODUCTION

DOUBLING AND HALVING

You will need: poster paper

Draw a Venn diagram on the board or on poster paper. Label one circle 'Doubling' and the other 'Halving'. Have students brainstorm ideas about doubling and halving and use the Venn Diagram to sort them. Include examples if relevant. Retain the diagram for the reflection component of the lesson.

INDEPENDENT TASKS

Note: Choose from Tasks 1, 2 or 3.

You will need: BLM 3 'Tables Chart 1', BLM 4 'Tables Chart 2' poster paper, scissors, *Excel*, *Word*, Student Book p. 61 'Doubling and Halving'

TASK 1: DOUBLING AND HALVING IN TABLES SETS

Give pairs of students copies of BLM 3 'Tables Chart 1' and BLM 4 'Tables Chart 2'. Have them select two tables sets, e.g. 2 × and 4 ×, and stick them on a piece of poster paper. Then have them identify all of the individual tables that are based on doubling and those based on halving. Have them decide how to record this. Repeat until three sets have been examined.

TASK 2: INTERACTIVE TASK

Have students use *Excel* and run the answers to the two times tables down the cells. Then have them double the answers in the next column, and then double again in the third column. Ask, 'What do you notice?' Repeat with the three, six and twelve times tables, then with the five and ten times tables. Have students print their findings and write comments on the bottom of the page, or have them paste their findings into a *Word* document and include comments.

TASK 3: STUDENT BOOK p. 61 *'Doubling and Halving'*

TEACHING GROUP

You will need: modelling equipment (e.g. counters, Unifix blocks)

MODELLING DOUBLING AND HALVING

- For students who require support, give them some modelling equipment, e.g. counters or Unifix blocks. Give students a number, and have them make the number with the modelling equipment. Then have them double or halve the number. Have students record the numbers and the results, indicating if it was doubling or halving. Repeat this a number of times.

DOUBLING LARGE NUMBERS

- For students who require a challenge, have them multiply large numbers by 4 and 8, by doubling and doubling and so on. Set a group of multiplications for them to complete, e.g. 4 × 26, 4 × 33, 8 × 126, 8 × 227. Challenge students not to use calculators.

REFLECTION

Select from the following to suit your class and their learning outcomes:

- Invite students to share the charts they made in Independent Tasks, Task 1. Examine how students have recorded or indicated the patterns.
- Have students share the large doubling equations they solved in Independent Tasks, Task 2.
- Revisit the Venn diagram constructed in the Whole-Class Introduction and add any further ideas or comments about doubling and halving.

LESSON PLAN 3

TUNING IN

PARTIALLY COVERED ARRAYS

You will need: NTO 4.10 'Arrays'

Show students a partially covered array. This can be generated using NTO 4.10 'Arrays'. Have students work out the total number of items in the array, including those not in view. Ask students to explain how they worked out their answers.

WHOLE-CLASS INTRODUCTION

HIDDEN AMOUNTS

You will need: NTO 4.10 'Arrays'

Generate arrays using NTO 4.10 'Arrays'. Discuss the importance of multiplication in finding the values of the arrays, whether they are visible or hidden. Revisit how an equation links to an array with a number of examples on the board, either by providing an equation and having students draw the array or providing an array and having students provide the equation. Use arrays with circles and grids.

INDEPENDENT TASKS

Note: Choose from Tasks 1, 2 or 3.

You will need: BLM 28 'Arrays 1', BLM 29 'Arrays 2', *PowerPoint*, Student Book p. 62 'Hidden Arrays'

TASK 1: CREATING PROBLEMS

Give students an array. These could be copied from BLM 28 'Arrays 1' and BLM 29 'Arrays 2'. Have students write the related equation, and then write a word problem that could be linked. Have students record their work. Repeat a number of times.

TASK 2: INTERACTIVE TASK

Once students have written a number of problems during the previous task, have them create a *PowerPoint* slide for each problem. If the array is required in the problem, the student could scan or create the array in *PowerPoint*. Ensure students can find the solution to their problem. Have them test their questions on a friend.

TASK 3: STUDENT BOOK p. 62 *'Hidden Arrays'*

TEACHING GROUP

You will need: modelling equipment (e.g. counters), a piece of card, LO: *L2055* or LO: *L2053 'Arrays: Word Problems'*

MODELLING HIDDEN ARRAYS

- For students who require support, have them use counters to create an array such as 4 × 5. Ask, 'How many counters? How many rows? How many columns?' Then cover part of the array with a piece of card. Ask, 'How many counters can you see? But how many counters are there in total?' Repeat with a different array. Then create and partially cover an array so students don't know how many counters there are. Discuss strategies for working out the total. Take the card away so students can check their answers.

LO 2055 AND LO 5053 'ARRAYS WORD PROBLEMS'

- For students who require a challenge, have them work independently on computers, using LO: *L2055* or LO: *L2053 'Arrays: Word Problems'* to create multiplication or division equations with missing numbers. Then students use their multiplication or division facts to find the missing number, with the support of arrays. Note: the concept of remainders is used in these Learning Objects.

REFLECTION

Select from the following to suit your class and their learning outcomes:

- Have students share the problems they created in Independent Tasks, Task 1, either electronically or on hard copy. Have other students attempt to solve the problems. Check answers.
- Have students reflect on the challenges of writing array problems in Independent Tasks, Task 2. Ask, 'What did you find difficult? What was easy? Was it hard to write a question everyone could understand? What was it like when the other students completed the question?'
- Using NTO 4.10 'Arrays', present a number of hidden arrays for students to work out. Ask students to explain their strategies.

Home Tasks

Select from the possible Home Tasks:

- Have students take home the array problems they created in Lesson Plan 3, Independent Tasks, Task 1, to try out on their parents or carers or siblings.
- Have students write a comment or reflection on how their array problems went at home. Have them include any suggested modifications.

Assessment

- Have students complete **Student Assessment p. 63**.
- Review with students **Assessment Task Card 4.15**.

During the three lessons:

- Make a copy of students' written reflections from Lesson Plan 3 on writing array problems.
- Collect copies of digital materials, such as the *Excel* materials in Lesson Plan 2, Independent Tasks, Task 2, to add to students' digital portfolios.
- Make note of students who are still struggling with basic concepts and spend time with them, reviewing basic facts.

Recommendations for Future Learning

Specific to Student Assessment p. 63; if the student is experiencing some difficulty:

Q 1 Review the concept of commutativity. Have the student use BLM 3 'Tables Chart 1' and BLM 4 'Tables Chart 2' to look for links. Have the student practise writing the relevant equation if given a single equation. For example: if provided with $3 \times 4 = 12$, the student should be able to write $4 \times 3 = 12$.

Q 2–3 Use modelling equipment, e.g. counters or Unifix blocks, to model doubling and halving. Work with even numbers for halving so there are no remainders. The student could be further supported with calculators.

Q 4 Review arrays and the link to multiplication equations. Use modelling equipment to explore arrays and what happens when parts of them are covered up.

If the student has not achieved the recommended skills for this unit:

1. See **Assessment Task Card 4.15** for specific recommendations.
2. Review basic tables facts. This could be supported with BLM 3 'Tables Chart 1' and BLM 4 'Tables Chart 2' or multiplication grids.
3. Use modelling equipment to support the concepts of arrays, doubling and halving.
4. Review *Nelson Maths: Australian Curriculum NSW Year 3* Units 18, 19 and 25.

If the student has achieved the recommended skills and these skills are firmly established, consider:

1. Having the student complete 'Commutative Property of Multiplication' and 'Halving' from *Nelson Maths Building Mental Strategies Big Book 4*, pp. 48–49, 64–65.
2. Moving forward to *Nelson Maths: Australian Curriculum NSW Year 5* Unit 7.
3. Extending the student in any of the listed activities by working with larger numbers, particularly for the doubling and halving activities.
4. Integrating the use of technology for the exploration of arrays and number line links in doubling and halving.

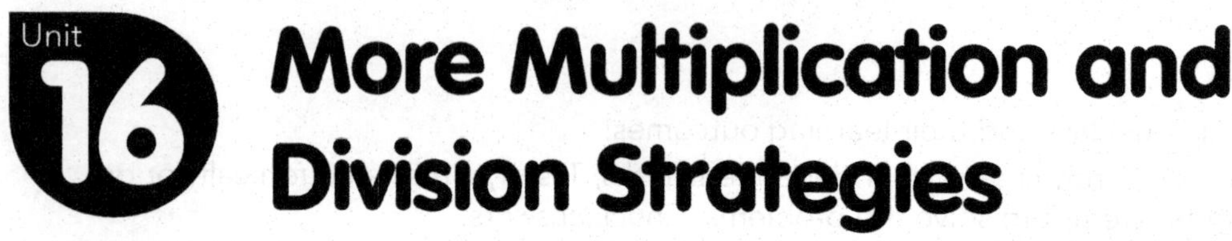

Unit 16 More Multiplication and Division Strategies

Number and Algebra

Multiplication and division MA2-6NA uses mental and informal written strategies for multiplication and division

division, remainder, doubling, factors, halving, multiplication, place value, strategy

LESSON PLAN 1

TUNING IN

MULTIPLICATION CHALLENGE

You will need: poster paper

Have groups of students work on a multiplication challenge. Give students a piece of poster paper and the numbers 16, 20 and 36, and have them write as many multiplication equations that equal those numbers as they can. Set a time limit, according to students' abilities. Share some of the equations at the completion of the activity. Retain the posters for the next activity.

WHOLE-CLASS INTRODUCTION

RELATED DIVISION

You will need: posters from Tuning In

The next challenge is to find as many division equations as possible related to the multiplication ones found in Tuning In and to write them on the poster. Ask, 'What do you notice about the multiplication and division equations for 16?' The answer is that they all use same numbers, i.e. 1, 16, 4, 8, 2. Tell students these are called factors. Have students identify the factors of 20 and 36.

INDEPENDENT TASKS

Note: Choose from Tasks 1, 2 or 3.

You will need: poster paper, BLM 3 'Tables Chart 1', BLM 4 'Tables Chart 2', LO: *L2060 'Arrays: explore factors'*, Student Book p. 64 'Exploring Arrays'

TASK 1: EXPLORING FACTORS

Give pairs of students poster paper. Have students repeat the activities from Tuning In and the Whole-Class Introduction for the numbers 45, 50, 24 and 48 (according to students' abilities). If required, students could be supported with BLM 3 'Tables Chart 1' and BLM 4 'Tables Chart 2'.

TASK 2: INTERACTIVE TASK

Have students work independently on computers, using LO: *L2060 'Arrays: explore factors'* to explore how numbers can be broken up with factors, using arrays. Note: there are various versions of this Learning Object to cater for different abilities.

TASK 3: STUDENT BOOK p. 64 *'Exploring Arrays'*

TEACHING GROUP

You will need: BLM 3 'Tables Chart 1', BLM 4 'Tables Chart 2', coloured highlighters

EQUATION HUNT

- For students who require support, give them BLM 3 'Tables Chart 1' and BLM 4 'Tables Chart 2' and some coloured highlighters. Ask students to find all of the multiplication equations that equal a target number, e.g. 48. Have them complete the hunt and then share answers. List all the equations that make that total, then have students identify the related division equations. Circle the key numbers to draw out the factors. Don't forget to include 48 × 1! Students can repeat the activity with a different target number and a different-coloured highlighter.

SQUARE NUMBERS

- For students who require a challenge, have them explore square numbers and draw the matching arrays. Students can also identify the factors associated with the square numbers. If students need further extension, they could explore triangular numbers, initially through drawing diagrams.

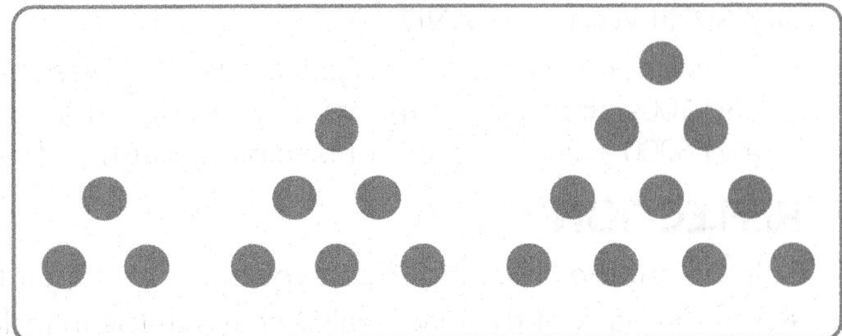

REFLECTION

Select from the following to suit your class and their learning outcomes:

- Have students share their posters of multiplication and division equations and factors from Tuning In and the Whole-Class Introduction. Check to see that all of the factors are included. Ask, 'How did you organise your work to check you had all of the equations?'
- Have students from the Teaching Group 'Equation Hunt' share their work.
- Give students a number, e.g. 100, and see if, as a group, they can find all of the factors.

LESSON PLAN 2

TUNING IN

PLACE-VALUE CHART AND MULTIPLICATION

You will need: NTO 4.6 'MAB: Thousands Mat'

Show a number, e.g. 4, on NTO 4.6 'MAB: Thousands Mat'. Ask, 'What is the answer if we multiply 4 by 10 – that is, 4 × 10? How could we show this on the place-value chart?' Discuss, and invite students to the board to show how they would use the chart to solve this. Ask, 'What if we multiplied 4 by 100?' Repeat with a variety of numbers.

WHOLE-CLASS INTRODUCTION

MULTIPLYING BY FACTORS OF 10

Simplify the place-value chart from Tuning In as shown at right. Record the multiplication of 4 by 10 and 100. Repeat with a number of different examples. To extend the activity, present students with 20 × 6 and ask, 'How could we work this out?' Accept students' ideas and record on the board. For example: 2 × 6 and then multiply by 10.

H	T	O	
		4	
	4	0	(4 × 10)
4	0	0	(4 × 100)

INDEPENDENT TASKS

Note: Choose from Tasks 1, 2 or 3.

You will need: poster paper, *Excel*, Student Book p. 65 'Multiplying by 10s and 100s'

TASK 1: MULTIPLYING BY 10 OR 100

Give students poster paper and tell them to multiply by 10 or 100 (approximately half the class for each, depending on students' abilities). Provide some starting numbers, e.g. 2, 6, 8, 9. For students needing extension, provide either numbers with 10s, e.g. 12, or multiplication equations, e.g. 3 × 20. Have students produce a place-value chart showing the relevant multiplications.

TASK 2: INTERACTIVE TASK

Have students use *Excel* to create electronic place-value charts. These could be similar to those made in the Whole-Class Introduction or, if suitable, students can generate simple formulas to complete the multiplication by 10 and 100. If using the formula, have students complete an exploration of multiplying by 10s and then, in separate columns, multiplying by 100s. Give students some starting numbers, such as 2, 7, 8 and 9, and then move to numbers such as 12 and 22. Extend students using the formula in *Excel* by giving them multiplication equations such as 2 × 40.

TASK 3: STUDENT BOOK p. 65 *'Multiplying by 10s and 100s'*

TEACHING GROUP

You will need: calculators

MULTIPLYING BY 10 AND 100 WITH CALCULATORS

- For students who require support, give them calculators to use to explore patterns, e.g. 2 × 3, 20 × 3, 200 × 3. Have students record the questions and the answers. Repeat a number of times with different examples. Have students articulate the patterns that they see.

DIVISION WITH 10S AND 100S

- For students who require a challenge, organise them into pairs to work backwards with the division of 10s and 100s. Start with numbers such as 500 ÷ 100, 500 ÷ 10. Then extend to multiples such as 6000 ÷ 200 and 6000 ÷ 20. Have students explore patterns based on tables facts.

REFLECTION

Select from the following to suit your class and their learning outcomes:

- On the back of the place-value charts made in Independent Tasks, Task 1, have students write a summary of what they found through their explorations of multiplying by 10 or 100. Invite students to share their summary with the class.
- Have students share the place-value charts made in Independent Tasks, Task 1, with the class. Examine groups of multiplications by 10 and look for commonalities. Repeat with multiplications by 100. Then compare one of each to again look at the commonalities and differences.
- Go through Student Book p. 65 'Multiplying by 10s and 100s' with students. Draw out any patterns students observe, e.g. if we multiply by 100, the number always ends in two zeros.

LESSON PLAN 3

TUNING IN

DIVISION WITH REMAINDERS

You will need: BLM 30 'Multiplication Grid'

Using BLM 30 'Multiplication Grid', have students work backwards, i.e. provide a number and have them find the two numbers that multiply to give that answer. Repeat a number of times. To extend the activity, give students an equation, e.g. 20 ÷ 5, and have students use the multiplication grid to find the answer.

WHOLE-CLASS INTRODUCTION

DIVISION

Review with students the different ways to write division equations, i.e. 20 ÷ 5 = 4 and $5\overline{)20}$ with 4 written above.

Go over each component of the equation and the representations, e.g. the answer goes at the top, don't forget to line up place-value columns. Write a number of equations on the board, based on tables facts, and have students offer answers.

INDEPENDENT TASKS

Note: Choose from Tasks 1, 2 or 3.

You will need: BLM 30 'Multiplication Grid', counters, LO: *L2008 'The divider: whole number remainders'*, Student Book p. 66 'Division with Remainders'

TASK 1: USING ARRAYS TO SHOW REMAINDERS

Show students 21 counters arranged randomly. Ask them to make an array by grouping them into threes. Establish the related multiplication and division equations (7 × 3 = 21, 3 × 7 = 21, 21 ÷ 3 = 7 and 21 ÷ 7 = 3). Rearrange the counters randomly. Ask them to make a different array by grouping them into twos. Discuss the 'leftover' counter and show that it can be written as a remainder. Model how the multiplication grid can also be used to solve the problem. Repeat with other numbers as the dividend.

TASK 2: INTERACTIVE TASK

Have students work independently on computers, using LO: *L2008 'The divider: whole number remainders'*, to solve whole-number division problems (involving remainders) using partitioning.

TASK 3: STUDENT BOOK p. 66 *'Division with Remainders'*

TEACHING GROUP

You will need: counters or blocks

PRACTISING DIVISION

- For students who require support, amend the activity in Independent Tasks, Task 1 to nine counters. Ask them to make an array by grouping them into threes. Establish the related multiplication and division equations (3 × 3 = 9, and 9 ÷ 3 = 3). Rearrange the counters randomly. Ask them to make a different array by grouping them into twos. Discuss the 'leftover' counter and show that it can be written as a remainder. Repeat using four and five as the divisor.

WRITING WORD EQUATIONS

- For students who require a challenge, give them an equation such as 5 × 9. Have students work out the answer, and the two respective division equations. Then students write a word problem to match the equation. For example: 'Helen had 45 chickens and 9 pens. How many chickens could she place evenly in each pen?' Amend the task so that the problem involves a division equation with remainders.

REFLECTION

Select from the following to suit your class and their learning outcomes:

- Have students share their equations from Independent Tasks, Task 1. Ask, 'Did you find there were numbers that came up more than once? Why do you think this may happen?'
- Have students from the Teaching Group 'Writing Word Equations' share the problems they wrote. Students ask the class their questions and then have them attempt to solve it.
- Ask, 'What do we do with remainders in real life? For example, how could we share three apples between two people?'

Home Tasks

Select from the possible Home Tasks:

- Have students take home the charts they made in Lesson Plan 2, Independent Tasks, Task 1, and explain to their parents or carers what they discovered.
- Have parents or carers practise tables with the student. BLM 3 'Tables Chart 1' and BLM 4 'Tables Chart 2' could be photocopied and sent home to help.

Assessment

- Have students complete **Student Assessment p. 67**.
- Review with students **Assessment Task Card 4.16**.

During the three lessons:

- Make a copy of students' written reflections from Lesson Plan 3 on their knowledge and confidence with tables.
- Collect copies of digital materials, such as the *Excel* charts from Lesson Plan 2, Independent Tasks, Task 2, to add to students' digital portfolios.
- Make note of tables students are struggling with and have these as a focus to practise at home.

Recommendations for Future Learning

Specific to Student Assessment p. 67; if the student is experiencing some difficulty:

Q 1 & 4 Review the concepts of multiplication and division. This could be supported with materials such as BLM 3 'Tables Chart 1' and BLM 4 'Tables Chart 2' or BLM 30 'Multiplication Grid', or modelling equipment such as Unifix blocks and counters.

Q 2–3 Review multiplication by 10 and 100, supported with the place-value chart. NTO 4.6 'MAB: Thousands Mat' could be used, along with modelling equipment on a hand-drawn place-value chart.

If the student has not achieved the recommended skills for this unit:

1. See **Assessment Task Card 4.16** for specific recommendations.
2. Review basic tables facts. This could be supported with BLM 3 'Tables Chart 1' and BLM 4 'Tables Chart 2' or BLM 30 'Multiplication Grid'.
3. Use modelling equipment to support the ideas of multiplying by 10 and 100. Calculators could also be used to develop ideas and patterns.
4. Review *Nelson Maths: Australian Curriculum NSW Year 3* Unit 18, 19 and 25.

If the student has achieved the recommended skills and these skills are firmly established, consider:

1. Having the student complete 'Multiplying by Multiples of 10', 'Multiplying by 100' and 'Factors' from *Nelson Maths Building Mental Strategies Big Book 4*, pp. 60–63, 68–69.
2. Moving forward to *Nelson Maths: Australian Curriculum NSW Year 5* Unit 10.
3. Extending the student in any of the listed activities by working with larger numbers, particularly for multiplying by 10 and 100.
4. Giving the student experience with word problems based on tables facts.

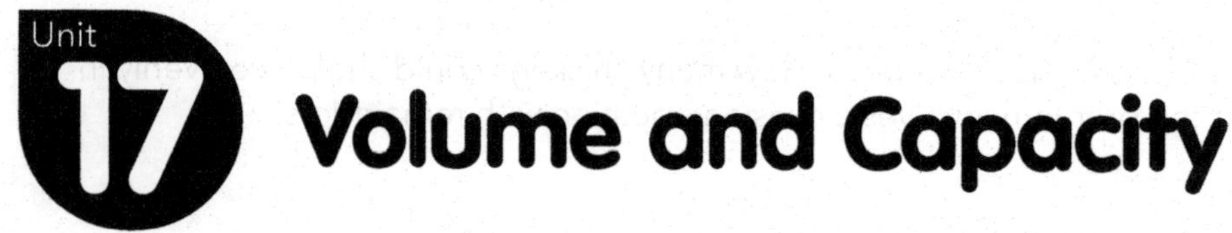

Unit 17 Volume and Capacity

Measurement and Geometry

Volume and capacity MA2-11MG measures, records, compares and estimates volumes and capacities using litres, millilitres and cubic centimetres

 cubes, measure, measuring unit, volume, cubic centimetres (cm^3), capacity, litres (L), millilitres (mL)

LESSON PLAN 1

TUNING IN

A LITRE OR NOT A LITRE

You will need: five different-sized containers and a litre jug

Show students five containers that do not have their capacities marked. Have two with a capacity of one litre but with a completely different appearance (e.g. a prism-shaped juice container and a round bottle). Label the containers A to E. Ask students individually, then in small groups, to decide whether each container contains less than, about or more than a litre. Ask students to justify their responses.

WHOLE-CLASS INTRODUCTION

LESS THAN A LITRE

You will need: a cup, a teaspoon and a litre jug

Following on from the Tuning In activity, establish that a cup and teaspoon each have a capacity of less than a litre. Ask how many cups full would be equivalent to a litre and establish the capacity of a cup as about a quarter of a litre. Discussion about repeating the activity with a teaspoon should lead to the conclusion that it would be time-consuming to find the capacity of a teaspoon as a fraction of a litre. This will conveniently lead to the need for a unit of measurement that is smaller than a litre: the millilitre.

INDEPENDENT TASKS

Note: Choose from Tasks 1, 2 or 3.

You will need: a variety of hands-on equipment (e.g. a graduated litre jug and containers smaller than a litre), LO: *L1994 'Squirt: two containers: Level 1'*, Student Book p. 68 'Smaller than a Litre'

TASK 1: INVESTIGATING CAPACITY

Give small groups a variety of containers and a graduated litre jug. Ask students to estimate the capacity of each container in terms of how many of each would be needed to fill the litre jug and the number of millilitres each holds. Following this, the students check their estimates by completing the activity.

TASK 2: INTERACTIVE TASK

Have students work independently on computers, using LO: *L1994 'Squirt: two containers: Level 1'*, to compare the capacities of two containers.

TASK 3: STUDENT BOOK p. 68 *'Smaller than a Litre'*

TEACHING GROUP

You will need: a litre jug, containers with capacities of 500 mL and 250 mL, containers of various sizes that do not have their capacities marked

HALF A LITRE, QUARTER OF A LITRE

- For students who require support, begin with comparing a litre with a half litre. Ask students to estimate the number of times the 500 mL container will need to be poured into the litre jug to fill it. Let them establish by using measuring equipment that two times half a litre is the same as one litre, and then that half a litre is the same as 500 mL. Repeat with a 250 mL container.

ACCURATE CAPACITY

- For students who require a challenge, have them identify accurate ways of finding the exact capacities of containers of various sizes by using different combinations of containers.

REFLECTION

Select from the following to suit your class and their learning outcomes:

- Ask, 'How could we work out if a container holds more or less than 1 litre?' Have students share their ideas.
- Ask, 'Why do we need a unit to measure capacity that is less than a litre?' Prompt discussion by showing students different sized containers.
- Ask students 'What is half a litre? What does it look like? How could we measure it?' Ask 'Is half a litre a formal measurement?' Have students share their responses.

LESSON PLAN 2

TUNING IN

HOW ELSE CAN WE SAY IT?

You will need: a marked 1.5 litre container or a picture of one

Review or introduce the term 'capacity'. Collect ideas on the board about what the term means. Show students the 1.5 litre container and ask, 'What are the different ways the capacity of the container be shown?' (1 500 mL, 1 L 500 mL, $1\frac{1}{2}$ litres).

WHOLE-CLASS INTRODUCTION

FIND THE SAME AMOUNT

You will need: six containers (or pictures of them) with capacities marked, three of which have the same capacity

Show six containers that have their capacities marked either in litres with a decimal, in litres and millilitres or in millilitres. The three that have the same capacity should have their capacities marked in three different ways (litres with a decimal, litres and millilitres, and in millilitres). The capacities of the containers might be, for example, 1250 mL, 1.2 L, 1 L 250 mL, 1300 mL, 1.25 L, 1L 205 mL. Tell the students that three of the capacities are the same and ask them, in small groups, to identify them.

INDEPENDENT TASKS

Note: Choose from Tasks 1, 2 or 3.

You will need: containers A, B, and C without capacities marked, graduated measuring jugs, water, Student Book p. 69 'Litres and Millilitres'

TASK 1: FIND THE CAPACITY

Ask students to find the capacities of three containers labelled A, B and C (e.g. 500 mL, 1.5 L and 2.25 L) using measuring equipment. Ask them to record the capacities of each in three ways (litres with a decimal, litres and millilitres, and in millilitres).

TASK 2: INTERACTIVE TASK

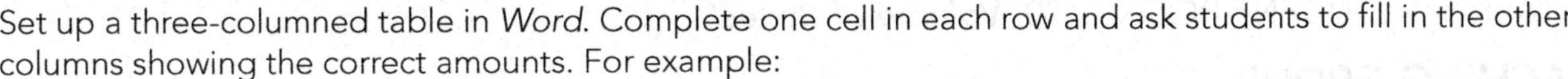

Set up a three-columned table in *Word*. Complete one cell in each row and ask students to fill in the other columns showing the correct amounts. For example:

Litres with a Decimal	Litres and Millilitres	Millilitres
2.5 L		
	1 L 200 mL	
		1 300 mL

TASK 3: STUDENT BOOK p. 69 *'Litres and Millilitres'*

TEACHING GROUP

You will need: 1 and 2 litre measuring jugs, various containers

CONVERTING BETWEEN UNITS OF CAPACITY

- For students who require support, begin by converting between two ways of showing capacity, e.g. have a marked 2 litre jug that includes millilitres, and ask students how many millilitres it contains. Have students read the measurements of different units off the jug. Revisit the link between whole litres and halves. As they increase in confidence, show a marked 1.5 L bottle of a drink and ask how many millilitres it contains.

WHAT IS THE ACTUAL CAPACITY

- Commercial drink containers usually hold slightly more than the stated amount that is in them. For students who require a challenge, ask them to find the actual, accurate capacity of one or more drink cartons.

REFLECTION

Select from the following to suit your class and their learning outcomes:

- Ask, 'Does the tallest container always have the biggest capacity?'
- Ask, 'Why do you think the capacity of one drink container is shown as 1.5 L and another as 1 L 500 mL?'
- Ask, 'Does a 1 litre milk carton hold exactly one litre? Why?' Invite students from the teaching group activity to share ideas and their findings.

LESSON PLAN 3

TUNING IN

WHAT DO WE MEAN BY VOLUME?

You will need: an empty tissue box or similar

Using the 'Think, Pair, Share' technique, show students a tissue box or similar and ask, 'What do we mean by the volume of the box?' Students reflect, then share their ideas with a partner and then with the larger group.

WHOLE-CLASS INTRODUCTION

WHY CUBIC CENTIMETRES?

Following on from the Tuning In activity, ask students to reflect on their previous work on volume and suggest why it is expressed in cubes and not, say, in spheres. Discuss. Ask, 'Why do we need a uniform unit such as a cubic centimetre (cm^3) to describe volume?' Remind students that 8 cm^3 should be referred to as 'eight cubic centimetres' and not 'eight centimetres cubed' or 'centimetre cubes'.

INDEPENDENT TASKS

Note: Choose from Tasks 1, 2 or 3.

You will need: interlocking centimetre cubes, drawings of small, box-shaped models made from interlocking cubes, NTO 4:19 'Multilink Blocks', Student Book p. 70 'Volume of Boxes'

TASK 1: BOX-SHAPED MODELS

Show students a rectangular prism that is 3 cm long by 2 cm wide by 1 cm high (pictorially or in actual interlocking centimetre cubes). Establish the volume as 6 cm^3. Show them a similar one with a second layer added and ask, 'What is the volume?' Ask, 'If we added another layer, what would the volume be?' Repeat for other, small box-shaped models, perhaps using drawings if it is felt students are ready for this.

TASK 2: INTERACTIVE TASK

Have students work with NTO 4:19 'Multilink Blocks' to copy prepared drawings of small, box-shaped models. Each virtual block can be a representation of an interlocking centimetre cube and the volume of each model identified.

TASK 3: STUDENT BOOK p. 70 *'Volume of Boxes'*

TEACHING GROUP

You will need: interlocking centimetre cubes, Student Book p. 70 'Volume of Boxes'

MAKING A BOX-SHAPED MODEL

- For students who require support, begin with showing a model made from cubes and ask students to make the same model and then identify the volume. Following this, add a second layer. Encourage students to give the volume before checking by counting the cubes.

A WIRE-FRAMED BOX

- For students who require a challenge, show a drawing of a wire-framed box that is 4 cm long by 2 cm wide by 4 cm high. Show a complete layer on the bottom layer only and a single cube on each of the other three layers (see diagram). Ask students to work out the volume of the box.

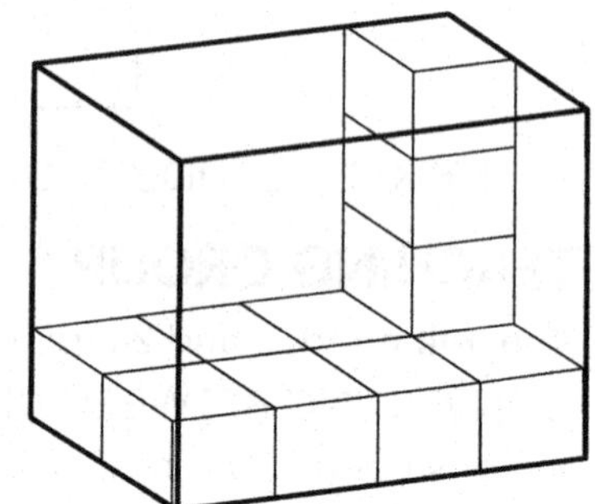

REFLECTION

Select from the following to suit your class and their learning outcomes:

- Ask, 'Why do we use cubes to show volume?'
- Show students an interlocking centimetre cube model that is 3 cm by 6 cm. Ask, 'What would the volume be if it were three layers high?'
- Ask, 'How is it possible to have two box-shaped models that look different but have the same volume?'

Home Tasks

Select from the possible Home Tasks:

- Ask students to find something at home that they think has a volume of 40 cm^3 or less. Ask them to justify their responses.
- Ask students to bring in clean containers that have their capacities marked on them.

Assessment

- Have students complete **Student Assessment p. 71**.
- Review with students **Assessment Task Card 4.17**.

During the three lessons:

- Look for areas of difficulty in converting between units of capacity.
- Take note of how accurate the students are when finding capacity in practical situations.
- Take note of the degree to which students are able to work in the abstract when working out the volume of a rectangular prism.

Recommendations for Future Learning

Specific to Student Assessment p. 71; if the student is experiencing some difficulty:

Q 1 Model the difference between a millilitre and a litre in a practical activity.

Q 2–3 Provide a lot of opportunities for hands-on experiences in finding capacity and converting between litres and millilitres.

Q 4 Provide a lot of opportunities for making actual models from interlocking cubes when finding the volume of rectangular prisms.

If the student has not achieved the recommended skills for this unit:

1. Make containers available which contain multiples of 500, 200 and 100 mL.
2. Look for opportunities in the classroom environment to find capacity expressed in both litres and millilitres.
3. Have the student continue to work with the interlocking centimetre cubes and making models.
4. Give the student sketches and the completed models to match together.
5. Review *Nelson Maths: Australian Curriculum NSW Year 3* Unit 30.

If the student has achieved the recommended skills and these skills are firmly established, consider:

1. Asking the student to find out about the kilolitre as a unit of capacity.
2. Asking the student to find a way to work out how much water a dripping tap would waste in a day.
3. Moving forward to *Nelson Maths: Australian Curriculum NSW Year 5* Unit 4.
4. Moving towards a more formal representation of volume, building on knowledge of perimeter and area.

Decimals to Two Decimal Places

Number and Algebra

Fractions and decimals MA2-7NA represents, models and compares commonly used fractions and decimals

decimal, decimal point, hundredths, larger, number line, order, place value, smaller, tenths, units, whole number, zero

LESSON PLAN 1

TUNING IN

ORDERING NUMBERS 1–1000

You will need: A4 paper, scissors

Give each student a sheet of A4 paper. Have students fold and cut the paper into quarters. On each quarter, have students write: their favourite number between 1 and 1000, a number less than 500, a number greater than 800 and a number between 600 and 700. Have students work in pairs to order the eight pieces of paper from smallest to largest. To extend the activity, students could work in groups of four. Note: this can be used as a pre-assessment activity.

WHOLE-CLASS INTRODUCTION

EXPLORING PLACE VALUE WITH DECIMAL NUMBERS

You will need: NTO 4.5 'Place-Value Mat: Millions'

Generate numbers with either one or two decimal places, and have students place the digits into the relevant place-value columns using NTO 4.5 'Place-Value Mat: Millions'. Ask, 'What are each of the columns called? What happens after the decimal point? What happens if there is a zero in the number?' The activity could be completed on the whiteboard, by drawing a place-value chart with columns from thousands to hundredths, and giving students numbers to write into the chart.

INDEPENDENT TASKS

Note: Choose from Tasks 1, 2 or 3.

You will need: dice (four for each pair – a different colour for each student in the pair), *Excel*, Student Book p. 72 'Place-Value Chart'

TASK 1: PLACE-VALUE CHART

Have pairs of students draw a place-value chart with columns from tens to hundredths. Give each pair four dice – a different colour for each student in the pair. Have each student roll their dice, and create the largest number possible. Now have the two students combine their numbers to make a decimal number – one student's dice represent the whole number, the other's represent the decimal. Finally, have each student record the number on their place-value chart. Students collect 10 numbers, then swap the roles of whole number and decimal. Extend this activity by using more dice to create larger whole numbers.

TASK 2: INTERACTIVE TASK

Have students work independently on computers to create a place-value chart in *Excel*. Give students a set of numbers, and have them insert these into the correct columns on the place-value chart.

TASK 3: STUDENT BOOK p. 72 *'Place-Value Chart'*

TEACHING GROUP

You will need: three dice for each student who requires support

2-DIGIT NUMBERS

- For students who require support, give them two dice and have them roll to create a 2-digit number. Have students make 10 different numbers. Ask students to create a place-value chart, with columns for tens and hundreds, and record their numbers on their charts. When students are confident, give them a third dice, and have them create 2-digit numbers with one decimal place and record their numbers on a place-value chart.

ORDER THE NUMBER

- For students who require a challenge, give them a set of 10 numbers with two decimal places to order from smallest to largest. Include numbers such as 3.20, 408.63 and 408.4. Have students discuss why different numbers are larger or smaller. Students create their own sets of 10 decimal numbers to order, and swap with a partner.

REFLECTION

Select from the following to suit your class and their learning outcomes:

- Ask, 'Why are zeros important in decimal numbers?' Create a class chart about the importance of the zero in place value. Have students record their different ideas and display.
- Using an IWB, have students display the charts created in Independent Tasks, Task 2, to share with the class.
- Revisit NTO 4.5 'Place-Value Mat: Millions', and provide students with 'tricky' numbers to add to the chart, e.g. 0.42, 1.06, 3.20, 180.40, 293.04. Discuss.

TUNING IN

GUESS THE NUMBER

You will need: sticky notes

Draw a place-value chart on the board with columns from hundreds to hundredths. On a sticky note, write a number that would fit into the place-value chart, containing different values for each digit, e.g. 327.14. Invite students to guess the whole number. On the chart, record with symbols: ✔ for correct digit, ✘ for incorrect digit and ☐ for a digit in the number but in the incorrect position. Students keep guessing, and try to beat you in a set number of rounds, e.g. 8. The student who guesses correctly selects the next number.

WHOLE-CLASS INTRODUCTION

READING AND WRITING DECIMAL NUMBERS

You will need: BLM 34 'Decimal Numbers and Words 1', BLM 35 'Decimal Numbers and Words 2'

Enlarge and cut up BLM 34 'Decimal Numbers and Words 1' and BLM 35 'Decimal Numbers and Words 2'. Give each student a card and have them silently move around the room to find their pair. Then have all students order themselves from smallest to largest. Do this in individual sets, i.e. all the numbers, then all the word forms. Ask, 'Are the orders the same?'

INDEPENDENT TASKS

Note: Choose from Tasks 1, 2 or 3.

You will need: BLM 34 'Decimal Numbers and Words 1', BLM 35 'Decimal Numbers and Words 2', BLM 43 'Symbols', *PowerPoint*, Student Book p. 73 'Matching Decimal Numbers and Words'

TASK 1: NUMBER CARDS

Give groups of students cards from BLM 34 'Decimal Numbers and Words 1' and BLM 35 'Decimal Numbers and Words 2', and the symbols <, > and = from BLM 43 'Symbols'. Have students create statements with the cards, e.g. 1.3 > One and three hundredths. When the group agrees, have students record their statement on a sheet of paper. Each statement needs to include one numeric card and one word card. Note: review the symbols < and > if necessary.

TASK 2: INTERACTIVE TASK

Have students work independently on computers to create a *PowerPoint* presentation that has matching decimal numbers and words on each slide. Students should be encouraged to make the slides interactive, e.g. the user has to guess the answer or select the options. Students could include relevant pictures.

TASK 3: STUDENT BOOK p. 73 *'Matching Decimal Numbers and Words'*

TEACHING GROUP

You will need: 20 blank cards for each student who requires a challenge

NUMERIC ORDER

- For students who require support, play a comprehension game. Read out decimal numbers (you could use BLM 34 'Decimal Numbers and Words 1' and BLM 35 'Decimal Numbers and Words 2'), and have students record them in their books in the numeric form. Once students are confident, have them record the numbers read out in word form. Provide an incentive, e.g. once a student has 10 correct answers, they become the reader.

DECIMAL MEMORY GAME

- For students who require a challenge, give them 20 cards and have them develop their own decimal 'Memory' game, e.g. where they write the decimal number and matching word form on individual cards. Once complete, pairs of students combine two sets and play the game. Early finishers from other activities could be invited to play.

REFLECTION

Select from the following to suit your class and their learning outcomes:

- Cut out cards from BLM 34 'Decimal Numbers and Words 1'. Invite three students to sit in front of the whiteboard, with their backs against it. Attach one card above each student's head so they can't see it, and have them take turns to guess the number, using Yes/No questions. For example: 'Does my number have two decimal places? Is my number greater than 10?' Play a number of rounds, with different students taking turns to guess.
- Invite students to share what they learned in the Teaching Groups. Students describe the activity and their discoveries. Ask, 'What was easy? What was challenging?'
- Create a chart of students' strategies for writing decimal numbers in words. What little tricks did they use? Display the chart in the classroom.

LESSON PLAN 3

TUNING IN

LET'S RUN

You will need: a space to run, tape measure/trundle wheel, stopwatches

Have students measure a distance of 100 m. Revise with them how to use a stopwatch. Have them time each other running the 100 m length, and make note of their individual times. Ask, 'How do we write down/record the time?'

WHOLE-CLASS INTRODUCTION

ORDERING DECIMAL NUMBERS

You will need: BLM 34 'Decimal Numbers and Words 1', BLM 35 'Decimal Numbers and Words 2'

Show students pairs of numbers and have them identify which is the largest decimal. (You could use BLM 34 'Decimal Numbers and Words 1' and BLM 35 'Decimal Numbers and Words 2' for this.) Play this as a game: show students 10 pairs of numbers, have them silently record the largest in each pair on a sheet of paper, and then find who has the most correct.

INDEPENDENT TASKS

Note: Choose from Tasks 1, 2 or 3.

You will need: BLM 34 'Decimal Numbers and Words 1', BLM 35 'Decimal Numbers and Words 2', long rolls of paper, glue, LO: *L2005 'Scale matters: hundreds'*, Student Book p. 74 'Ordering Decimal Numbers'

TASK 1: DECIMAL NUMBERS AND WORDS

Give pairs of students the cards from BLM 34 'Decimal Numbers and Words 1' and BLM 35 'Decimal Numbers and Words 2' and a long roll of paper. Have students create a giant number line to scale, attaching the decimal numbers in the correct order. Leave the task fairly open-ended. Note: this could be used as informal assessment.

TASK 2: INTERACTIVE TASK

Have students work independently on computers, using a hundredths scale on LO: *L2005 'Scale matters: hundreds'* to locate or place numbers on a number line.

TASK 3: STUDENT BOOK p. 74 *'Ordering Decimal Numbers'*

TEACHING GROUP

You will need: BLM 34 'Decimal Numbers and Words 1', BLM 35 'Decimal Numbers and Words 2', long rolls of paper, times from a sporting event, poster paper

SMALLEST TO LARGEST

- For students who require support, use level-appropriate numbers from BLM 34 'Decimal Numbers and Words 1' and BLM 35 'Decimal Numbers and Words 2'. Have students work as a group to order the decimal numbers from smallest to largest, and then to create a number line. Give support and guidance through questioning as required.

QUICKEST TO SLOWEST

- For students who require a challenge, give them a set of times (out of order) from a sporting event, e.g. running, cycling, swimming, car racing. Have students place them into order from quickest to slowest. Have students present this task as a poster. Students could include pictures related to the sport.

REFLECTION

Select from the following to suit your class and their learning outcomes:

- Have students present the number lines they made in Independent Tasks, Task 1. Ask, 'Why did you place this number first? How did you decide your order? What was tricky about this task?'
- Have students write their running times from Tuning In on a piece of card. Attach these to the whiteboard. Then, as a group, have students order the times from quickest to slowest.

Home Tasks

Select from the possible Home Tasks:

- Have students look for things around the home that show numbers with two decimal places, e.g. items from the pantry, books, newspapers, measuring equipment. Have them list the item and the decimal.
- Have students look at a newspaper (hard copy or online) for examples of decimal numbers, e.g. results of sporting events, news reports or advertisements. Have students cut out/print/save and bring these to class to share. These items could be used in some of the activities in Lesson Plan 3.

Assessment

- Have students complete **Student Assessment p. 75**.
- Review with students **Assessment Task Card 4.18**.

During the three lessons:

- Collect created items such as the *Excel* place-value charts from Lesson Plan 1, Independent Tasks, Task 2, as work samples for students' portfolios.
- Make note of students who completed the scaffolding tasks or the more challenging activities in the Teaching Groups.
- Review the Student Book pages and note areas of difficulty.

Recommendations for Future Learning

Specific to Student Assessment p. 75; if the student is experiencing some difficulty:

Q 1 Work with whole numbers, then numbers with only one decimal place.

Q 2–3 Use BLM 34 'Decimal Numbers and Words 1' and BLM 35 'Decimal Numbers and Words 2' to revisit naming activities.

Q 4–8 Practise drawing number lines with whole numbers, before moving to numbers with one decimal place. Give the student numbers such as 3.1, 5.6 and 7.2.

If the student has not achieved the recommended skills for this unit:

1. See **Assessment Task Card 4.18** for specific recommendations.
2. Have the student work with whole numbers in any of the listed activities (at their level) before moving to decimal numbers. Scaffold the student with the use of the place-value chart to reinforce the values and positioning of the decimal parts.
3. Review *Nelson Maths: Australian Curriculum NSW Year 3* Unit 28.

If the student has achieved the recommended skills and these skills are firmly established, consider:

1. Moving forward to *Nelson Maths: Australian Curriculum NSW Year 5* Unit 18.
2. Having the student complete *Nelson Maths Mental Strategies Big Book 4*, pp. 8–9, to reinforce mental strategies with decimal numbers.
3. Extending the student in any of the listed activities by using decimal numbers to three decimal places.

Chance

Statistics and Probability

Chance MA2-19SP describes and compares chance events in social and experimental contexts

certain, chance, even chance, events, impossible, least likely, likely, most likely, possible, probability, unlikely

LESSON PLAN 1

TUNING IN

ORDERING CHANCE WORDS

You will need: BLM 36 'Chance Words'

Enlarge, photocopy and cut up BLM 36 'Chance Words'. As a class, discuss and order the cards, perhaps across the front of the classroom. When complete, have students think of examples for each word.

WHOLE-CLASS INTRODUCTION

SPINNERS

You will need: NTO 4.12 'Spinners'

Building on the language from the Tuning In activity, have students predict the chance of landing on a particular colour on one of the spinners. Have students spin the spinner at least 10 times to test their theories. Repeat with the different spinners and have students draw conclusions, using the language of chance.

INDEPENDENT TASKS

Note: Choose from Tasks 1, 2 or 3.

You will need: BLM 36 'Chance Words', BLM 37 'Circles Template', coloured pencils/felt pens, rulers, LO: *L115 'The slushy sludger: questions'*, Student Book p. 76 'Chance in Everyday Life'

TASK 1: MAKING SPINNERS

Give students a copy of BLM 36 'Chance Words' and BLM 37 'Circles Template'. Have students design and create spinners that match words from BLM 36 'Chance Words' (these can be selected by you or the student). Have students write how the spinner reflects the terms selected (this can be done on the back or front of the spinner). This activity could be extended by having students make another spinner from the template.

TASK 2: INTERACTIVE TASK

Have students work independently on computers, using LO: *L115 'The slushy sludger: questions'* to operate a vending machine to squirt coloured 'slushies' into ice-cream cones, working out which 'sludge events' are possible and then choosing a matching probability word.

TASK 3: STUDENT BOOK p. 76 *'Chance in Everyday Life'*

TEACHING GROUP

You will need: BLM 36 'Chance Words', LO: *L2380 'Spinners: explore'*

WORKING ON LANGUAGE

- For students who require support, give them a copy of BLM 36 'Chance Words'. As a group, discuss the terms and make sure that students understand what each term means. Then have students stick the words in order on a sheet of paper, and write or draw an example next to each term.

LO: L2380 'SPINNERS: EXPLORE'

- For students who require a challenge, have them work in pairs, using computers to explore LO: *L2380 'Spinners: explore'* to test a coloured spinner with three equal-sized sectors. Students can also build and explore their own spinners.

REFLECTION

Select from the following to suit your class and their learning outcomes:

- Have students share the spinners they made in Independent Tasks, Task 1. Students could swap the spinners and test them, using the chance words to describe the outcomes.
- Have students discuss their experiences in Independent Tasks, Task 2. Ask, 'What language was used?'
- Give students some scenarios, e.g. 'We are going to lunch soon', and have them use the appropriate chance terminology.

LESSON PLAN 2

TUNING IN

ONE CANNOT HAPPEN IF THE OTHER HAPPENS

You will need: sticky notes

Ask, 'What is the weather like today? If it is wet, can it be dry at the same time?' Discuss. Investigate other scenarios, e.g. If the door is open, can it be shut? If the tap is on, can it be off?

WHOLE-CLASS INTRODUCTION

TOSSING A COIN

You will need: NTO 4.13 'Coin Flip', NTO 4.2 'Dice'

Using NTO 4.13 'Coin Flip', ask students to predict if a head or tail will be tossed, and then toss the coin. Ask, 'If a head is tossed, can we have a tail at the same time?' Discuss as a group. Repeat a number of times. Then use NTO 4.2 'Dice' to explore the event of one result happening, meaning the others cannot.

INDEPENDENT TASKS

Note: Choose from Tasks 1, 2 or 3.

You will need: blank cards, LO: *L116 'The slushy sludger: best guess'*, Student Book p. 77 'What's the Chance?'

TASK 1: OUR EXAMPLES

Give pairs of students some blank cards and have them develop a set number of examples of situations where one event cannot happen if the other happens. Have them write each example on a different card. They could share with another pair to make sure their ideas are clear.

TASK 2: INTERACTIVE TASK 

Have students work independently on computers, using LO: *L116 'The slushy sludger: best guess'* to operate a vending machine to squirt 'slushies' into ice-cream cones. Students investigate the chance of particular colours being served.

TASK 3: STUDENT BOOK p. 77 *'What's the Chance?'*

TEACHING GROUP

You will need: one dice for each student who requires support, coins, different-coloured counters in a cloth bag, LO: *L214 'The foul food maker: best guess'*

ROLLING DICE

- For students who require support, give them a dice. Have them predict and record which number they will roll. Students roll the dice and compare answers. Ask, 'If you roll a 6, can you get any other number?' Repeat the activity with coin tossing, and drawing different-coloured counters out of a bag.

LO: L214 'THE FOUL FOOD MAKER: BEST GUESS'

- For students who require a challenge, have them work on computers, using LO: *L214 'The foul food maker: best guess'* to work out the chance of getting a selection of an unpleasant meal.

REFLECTION

Select from the following to suit your class and their learning outcomes:

- Have students share their examples from Independent Tasks, Task 1. Discuss the chance of these events happening using the language from Lesson Plan 1. Display this in the room.
- Discuss students' results and findings from Student Book p. 77 'What's the Chance?' Ask questions, such as, 'Is the distance between the lines important? How did you test your ideas?'
- Have students reflect on what they discovered through the Learning Object activities. Ask, 'What was tricky? What was easy? How was chance important?'

TUNING IN

DICE ROLLING

You will need: NTO 4.2 'Dice' or real dice

Using NTO 4.2 'Dice' or real dice, ask students to predict the number that will be rolled. Roll the dice. Have students comment on their predictions. Ask students to predict again. Ask, 'Have you changed your prediction? Why?' Repeat a number of times.

WHOLE-CLASS INTRODUCTION

PREVIOUS RESULTS

You will need: NTO 4.13 'Coin Flip'

Using NTO 4.13 'Coin Flip', have students predict the outcome of tossing a coin. Test predictions. Ask students to predict again. Ask, 'Have you changed your prediction? Why?' Spend time talking about the chance of one thing happening that is not dependent on another or previous event, e.g. each time a coin is tossed, the chance of getting a head or tail is still the same.

INDEPENDENT TASKS

Note: Choose from Tasks 1, 2 or 3.

You will need: sets of playing cards, LO: *L168 'The vile vendor: best guess'*, Student Book p. 78 'Five Counters in a Bag'

TASK 1: EXPLORING PLAYING CARDS

Give pairs of students or small groups a set of playing cards. The task is to explore the chance of obtaining a 7 of Hearts in a certain number of draws, when replacing the card each time.

TASK 2: INTERACTIVE TASK

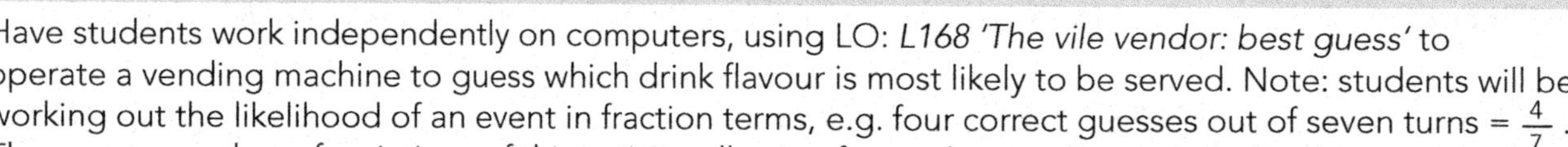

Have students work independently on computers, using LO: *L168 'The vile vendor: best guess'* to operate a vending machine to guess which drink flavour is most likely to be served. Note: students will be working out the likelihood of an event in fraction terms, e.g. four correct guesses out of seven turns = $\frac{4}{7}$. There are a number of variations of this activity, allowing for students to be extended.

TASK 3: STUDENT BOOK p. 78 *'Five Counters in a Bag'*

TEACHING GROUP

You will need: different-coloured counters, cloth bag, sets of playing cards

EXPLORING COUNTERS

- For students who require support, give them 10 counters. Talk about the chance of selecting a certain colour counter from a bag, e.g. a 2 in 10 chance of getting a red. Place the counters in a cloth bag. Have students draw out and replace a counter 10 times and record their result. Discuss what they found. Repeat, with students drawing out and not replacing 10 times. Discuss what they found. Ask, 'Did the results change? What was different about the second activity? Was the chance of drawing a red counter different? Why?'

PLAYING CARDS NOT REPLACED

- For students who require a challenge, give them sets of playing cards and have them explore the chances of obtaining a 7 of Hearts in a certain a number of draws, when not replacing the card each time.

REFLECTION

Select from the following to suit your class and their learning outcomes:

- Invite students to share what they discovered with the playing cards. Have students who completed Independent Tasks, Task 1, go first. Then move to students from the Teaching Group 'Playing Cards Not Replaced'. Ask, 'How are these two activities different?'
- Have students share their results from the Student Book p. 78 'Five Counters in a Bag'. What did they discover through completing the activity? Ask, 'Was selecting the counter the 30th time affected by the first draw? Why?'
- Have students suggest examples of events where the chance of one is not affected by the occurrence of the other, e.g. the chance of having a baby that is a boy or a girl is not affected by the sex of the previous baby.

Home Tasks

Select from the possible Home Tasks:

- Have students talk to parents or carers and come up with at least two examples for each of the following scenarios, and bring them to school and share:
  - Identify everyday events where one cannot happen if the other happens, e.g. something cannot be dry and wet at the same time.
  - Identify events where the chance of one will not be affected by the occurrence of the other, e.g. the chance of tossing a head or tail on a coin is not affected by the previous toss of the coin.

Assessment

- Have students complete **Student Assessment p. 79**.
- Review with students **Assessment Task Card 4.19**.

During the three lessons:

- Collect examples of students' spinners and the matching comments from Lesson Plan 1, Independent Tasks, Task 1, as an example of students' understanding of chance language.
- Make observations of students completing the interactive tasks. Are students engaged? What are they struggling with? Are they making the connections?
- Make a note of students who completed support or extension activities. Make note of any achievements or difficulties students experienced during these tasks.

Recommendations for Future Learning

Specific to Student Assessment p. 79; if the student is experiencing some difficulty:

Q 1–2 Review chance terminology, perhaps using BLM 36 'Chance Words'. These words could be attached next to events on the board and then ordered from least likely to most likely.

Q 3 Review chance situations that are everyday events, where one cannot happen if the other happens. Revisit the examples created by students both in class and as home tasks.

Q 4 Link the counters or coins activities to real-life examples. Discuss the possibilities of events happening, and then how they affect the next event.

If the student has not achieved the recommended skills for this unit:

1. See **Assessment Task Card 4.19** for specific recommendations.
2. Review chance terminology. Display it around the room, so the student can refer to it.
3. Continue to develop relevant examples of everyday events where one cannot happen if the other happens, and those where the chance of one happening will not be affected by the occurrence of the other.
4. Review *Nelson Maths: Australian Curriculum NSW Year 3* Unit 20.

If the student has achieved the recommended skills and these skills are firmly established, consider:

1. Moving forward to *Nelson Maths: Australian Curriculum NSW Year 5* Unit 19.
2. Having the student work with more complex examples.
3. Having the student numerically identify examples of chance, using either fractions or decimals, e.g. the chance of rolling a 4 on a dice is $\frac{1}{6}$.

Unit 20 Collecting Data

Statistics and Probability

Data MA2-18SP selects appropriate methods to collect data, and constructs, compares, interprets and evaluates data displays, including tables, picture graphs and column graphs

data, recording sheets, survey, tables, tally charts, two-way tables, Venn diagram

LESSON PLAN 1

TUNING IN

OUR PENCIL DATA

You will need: coloured pencils

Have each student choose their favourite coloured pencil. As a group, collect the data on the board, showing students' favourite coloured pencils. Allow students to guide this, to determine their prior knowledge in this area. Students may create a list, table or graph. To vary this activity, students could work in groups and then share their ideas with the whole class.

WHOLE-CLASS INTRODUCTION

COLLECTING OUR DATA

You will need: coloured pencils

Have each student now choose two pencils in their favourite colours, e.g. one red and one blue. As a group, create a list on the board of the colours chosen by each student, e.g. blue, green, blue, red (note: this will include multiples of some colours). Discuss what students notice. Ask, 'Is this an effective way of collecting data?' Then create a list of symbols, e.g. B for blue, or shade some of each colour on the board. Ask, 'Is this a more efficient way of collecting data?' Then work as a group to present the data as a table, perhaps vertical and/or horizontal (if students are ready), discussing the importance of labelling columns and rows. The tables may include numeric totals. Ask, 'Which method of data collection do you prefer, and why?'

INDEPENDENT TASKS

Note: Choose from Tasks 1, 2 or 3.

You will need: *Word* or *Excel*, Student Book p. 80 'Daisy's Data'

TASK 1: USING LISTS AND TABLES

Have pairs of students devise a question, e.g. 'What is your favourite football team?' Students collect data from their classmates, using their preferred method. When they have completed the collection, have them write two statements about what they discovered from their survey question.

TASK 2: INTERACTIVE TASK

Have students work in pairs on computers, using *Word* or *Excel* to collect data. Students devise a survey question, then survey the class and collect data electronically. Allow students to select their preferred method. Have them write two sentences below the data about what they discovered from their survey question.

TASK 3: STUDENT BOOK p. 80 *'Daisy's Data'*

TEACHING GROUP

You will need: a variety of hands-on materials, e.g. counters, teddies, blocks

CREATING GROUPS

- For students who require support, give them a selection of hands-on materials, e.g. counters, teddies, blocks. Students sort the materials into different categories, e.g. types of materials, colours. They record the data on a list or table of their choosing.

COLLECTING DATA ABOUT A PROBLEM

- For students who require a challenge, present them with a problem, e.g. finding out how much paper is thrown away in the classroom each day. Have students create and design a process for collecting data about the problem. Allow them to implement their process to examine its effectiveness.

REFLECTION

Select from the following to suit your class and their learning outcomes:

- Have students share their results from the survey they conducted in Independent Tasks, Task 1. Ask, 'What did you discover? Why did you decide to collect the data using that method?'
- Have students share the data they collected in Independent Tasks, Task 2. Ask, 'What did you discover? Why did you decide to collect the data using that method? Did using the computer make some things easier or harder? What were they?'
- Have students from the Teaching Group 'Collecting Data About a Problem' explain how they went about designing the process and then collecting data. Students present their findings to the group.

LESSON PLAN 2

TUNING IN

VENN DIAGRAMS

You will need: sticky notes

Give each student a sticky note to write their name on. Draw a Venn diagram on the board. Write 'Cat' above one circle and 'Dog' above the other. Then invite students to place their name on the diagram, depending on whether they have a cat or a dog. Ask, 'Where do I put my sticky note if I don't have a cat or a dog?' Count the sticky notes and write the numbers in each element of the Venn diagram. Keep the Venn diagram on the board. This activity can be varied by using hoops on the ground and having students stand in the hoops.

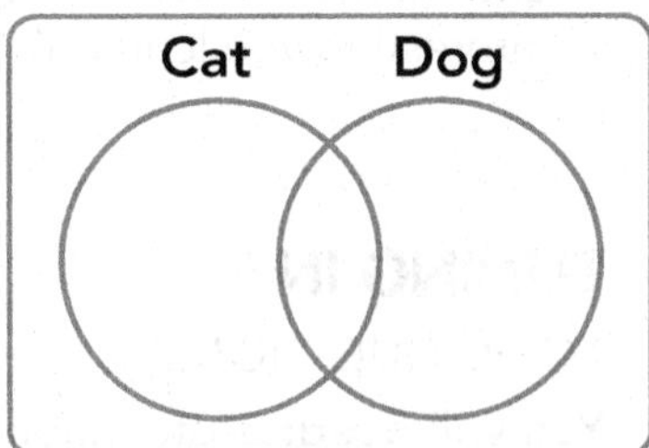

WHOLE-CLASS INTRODUCTION

2-WAY TABLES

You will need: sticky notes

	Has a dog	Doesn't have a dog
Has a cat		
Doesn't have a cat		

Repeat the Tuning In activity, with students writing their names on their sticky notes. This time, draw a 2-way table on the board. Have students add their sticky note to the relevant cell on the table. Count the sticky notes and write the numbers in each element of the 2-way table. Have students make statements, e.g. 'Thirteen people have a dog in our class'. Make the links between the Venn diagram and the 2-way table.

INDEPENDENT TASKS

Note: Choose from Tasks 1, 2 or 3.

You will need: *Word*, Student Book p. 81 'Venn Diagrams and 2-Way Tables'

TASK 1: SURVEYING FOOD

Have pairs of students develop a survey question about food and decide on the method of collection – either a Venn diagram or a 2-way table. Remind students they need two elements to their question. Have students collect the data and complete the presentation. Students write three sentences below their diagram/table about what they found. To extend the activity, students could convert the first set of data into the other format.

TASK 2: INTERACTIVE TASK

Have students work independently to create a question about music, and survey their classmates. Then students work on computers, using *Word* to collect their data into either a Venn diagram or a 2-way table. Students write three sentences below their diagram/table about what they found. Their information should be contained in a one-page document.

TASK 3: STUDENT BOOK p. 81 *'Venn Diagrams and 2-Way Tables'*

TEACHING GROUP

You will need: BLM 38 'Venn Diagrams'

VENN DIAGRAMS: GAMES

- For students who require support, give them BLM 38 'Venn Diagrams'. Have them write four questions about the top Venn diagram, e.g. 'How many people owned a 3DS and a Wii?' Students swap their questions for a partner to complete. Repeat with the second diagram.

VENN DIAGRAMS WITH THREE CIRCLES

- For students who require a challenge, have them develop and collect information that would require three circles on a Venn diagram. Students could look at Student Book p. 81 'Venn Diagrams and 2-Way Tables' or BLM 38 'Venn Diagrams' for ideas.

REFLECTION

Select from the following to suit your class and their learning outcomes:

- Have students share their findings about food from Independent Tasks, Task 1. Ask, 'Why did you decide to use a Venn diagram/2-way table? What did you learn from your survey?'
- Have students share their findings about music from Independent Tasks, Task 2. Ask, 'Why did you decide to use a Venn diagram/2-way table? What did you learn from your survey?' Have students reflect on the differences between using pen and paper and computers to collect their data.
- Have students from the Teaching Group 'Venn Diagrams: Games' share the questions they developed about games.
- Invite students from the Teaching Group 'Venn Diagrams with Three Circles' to share their diagrams and the questions they developed. Ask, 'How did you come up with the question? What was something interesting you learned?'

LESSON PLAN 3

TUNING IN

TOPIC QUESTIONS

You will need: poster paper

Give students a topic – it could be related to an issue at school, e.g. recycling or the school garden. Have small groups of students brainstorm three questions about the topic that they could collect data on, and record them on poster paper.

WHOLE-CLASS INTRODUCTION

SELECTING TOPIC QUESTIONS

You will need: questions on poster paper from Tuning In

Have students present their questions from Tuning In to the group. Discuss which questions were interesting, which questions would allow the collection of data and which questions could be reworded slightly to allow the collection of information. As a group, decide on five main questions.

INDEPENDENT TASKS

Note: Choose from Tasks 1, 2 or 3.

You will need: *Excel* and *Word*, poster paper, Student Book p. 82 'Our Question: My Report'

TASK 1: COLLECTING THE DATA

Divide the class into five groups and give each group a question from the Whole-Class Introduction activity. Have each group research and collect data about their question, presenting it on poster paper in one of the ways examined in the previous lessons, e.g. tables, Venn diagrams, 2-way tables. Students could visit other classes to collect the data.

TASK 2: INTERACTIVE TASK

Have students work as a group on computers to create a report of their collected data. This could include using *Excel* or *Word* to present the data and diagrams. Have students write interesting facts about their data.

TASK 3: STUDENT BOOK p. 82 *'Our Question: My Report'*

TEACHING GROUP

You will need: sample data (e.g. from a newspaper)

PRACTISING QUESTIONS

- For students who require support, provide them with a topic, e.g. the environment or water issues. Have each student write a question that could be surveyed about the topic. As a group, examine the question that would be the most effective. If there is time, have students survey and collect data from the rest of the class on the question, then, as a group, collate the data to determine the results.

INTERPRETING DATA

- For students who require a challenge, give them a set of data, e.g. from a newspaper. Have students determine what the data represents, then have them write three statements about what the data shows.

REFLECTION

Select from the following to suit your class and their learning outcomes:

- Have groups of students present their findings from Independent Tasks, Tasks 1 and 2. This could include paper materials and their electronic report. Have them share their interesting facts and findings.
- As a class, draw all the different questions from Independent Tasks, Tasks 1 and 2, together and make some conclusions about the investigation.
- Invite students to reflect on the activity. They may wish to use some of the ideas from Student Book p. 82 'Our Question: My Report'.

Home Tasks

Select from the possible Home Tasks:

- As a class, develop a question that students would like to survey parents and carers on. Have each student take the question home, survey the appropriate people and collect the data. Have students return to school with collected data, collate it and create a brief report. Have students report the results at home.

Assessment

- Have students complete **Student Assessment p. 83**.
- Review with students **Assessment Task Card 4.20**.

During the three lessons:

- Collect examples of students' questions and collected data, with their comments about what they learned, as examples of their ability to collect and interpret data.
- Collect examples of students' electronically created tables, Venn diagrams and 2-way tables to add to their digital portfolios, as evidence of their data collection and ability to interpret the findings.
- Collect students' written reflections about their learning from Lesson Plan 3.

Recommendations for Future Learning

Specific to Student Assessment p. 83; if the student is experiencing some difficulty:

Q 1 Review what a Venn diagram is. Show the student an example and practise interpreting data. Then have the student create their own question. Finally, have the student collect the data and create the representation on the Venn diagram.

Q 2 Review what a 2-way table is. Show the student an example and practise interpreting data. Then have the student create their own question. Finally, have the student collect the data and create the representation on a 2-way table.

Q 3 Have the student practise writing survey questions linked to sets of data, e.g. from a newspaper, then have them write their own questions.

If the student has not achieved the recommended skills for this unit:

1. See **Assessment Task Card 4.20** for specific recommendations.
2. Review data terminology and basic structures such as Venn diagrams and 2-way tables. Give the student templates to use.
3. Work with small data sets before moving to large data sets.
4. Allow the student to use relevant and available technology (e.g. *Word*, *Excel*) to support them with data organisation and presentation.
5. Review *Nelson Maths: Australian Curriculum NSW Year 3* Unit 11.

If the student has achieved the recommended skills and these skills are firmly established, consider:

1. Moving forward to *Nelson Maths: Australian Curriculum NSW Year 5* Unit 20.
2. Having the student work with larger data sets, e.g. the school population.
3. Having the student work with data sets from the media, and interpret the data or re-interpret it in different formats and presentations.

Unit 21 Number Patterns

Number and Algebra

Patterns and algebra MA2-8NA generalises properties of odd and even numbers, generates number patterns, and completes simple number sentences by calculating missing values

ML addition, multiplication, numbers, patterns

LESSON PLAN 1

TUNING IN

COUNTING PATTERNS

Present some counting patterns to students and have them continue the patterns. These may be based on tables or games such as 'Buzz'. Have students identify the starting number, the end point and the pattern. More complex patterns, e.g. add 10 subtract 3, could be used to extend students.

WHOLE-CLASS INTRODUCTION

ON CHARTS AND TABLES

You will need: NTO 4.1 'Hundred Chart', NTO 4.4 'Number Line: Patterns'

Display NTO 4.1 'Hundred Chart' and examine a range of number patterns with students. Look at starting points, end points, what happens with unit values in the patterns and how to record this information. Repeat with NTO 4.4 'Number Line: Patterns' and ask, 'Are the patterns the same or different? Why?' Gather ideas and record these on the whiteboard.

INDEPENDENT TASKS

Note: Choose from Tasks 1, 2 or 3.

You will need: BLM 1 'Hundred Chart', BLM 30 'Multiplication Grid', LO: *L3527 'Function machine'*, Student Book p. 84 'Number Patterns'

TASK 1: FINDING AND RECORDING NUMBER PATTERNS

Give some pairs of students BLM 1 'Hundred Chart' and others BLM 30 'Multiplication Grid'. Have pairs of students look for all the different patterns they can find on their chart. Have them record this. Students may wish to use a colour code or write descriptions of their patterns. Give students a time limit.

TASK 2: INTERACTIVE TASK

Have students work in pairs on computers, using LO: *L3527 'Function machine'* to feed the numbers 1, 2, 3 and 4 into a function machine, which automatically creates the first four numbers of a new number pattern. Students need to recognise the pattern and apply it.

TASK 3: STUDENT BOOK p. 84 *'Number Patterns'*

TEACHING GROUP

You will need: LO: *L589 'Musical number patterns: music maker'*

CREATING GROUPS

- For students who require support, have them work on computers, using LO: *L589 'Musical number patterns: music maker'* to make music by building up rhythms for four instruments. They choose a starting point on a number line and build a counting rule and create patterns. Note: there are various versions of this Learning Object to cater for different abilities.

MORE COMPLEX PATTERNS

- For students who require a challenge, present them with number patterns that use larger numbers and more complex sequences. Have students solve the sequence and then continue the patterns.

REFLECTION

Select from the following to suit your class and their learning outcomes:

- Have students share the patterns they located in Independent Tasks, Task 1. Look for similarities and for different patterns students have identified. Collect the ideas and display.
- Have students share their musical number patterns from Independent Tasks, Task 2. Invite students to explain how they identified the patterns and how they created the different elements.
- Present some number patterns on the board and have students identify and continue the patterns. You could include patterns with decreasing numbers. Students could be extended by considering patterns with fractions.

LESSON PLAN 2

TUNING IN

DOUBLING

Starting with 2, have students continue to double as many times as they can (without a calculator): 2, 4, 8, 16, 32, 64, 128 and so on. List the numbers in the pattern and ask students to make any observations about the numbers, e.g. the units 2, 4, 8, 6 continually repeat. Gather ideas and collect these on the whiteboard. Repeat with a different starting number, e.g. 3, and see if any observations hold true or if students can add to their comments.

WHOLE-CLASS INTRODUCTION

PATTERNS WITH MULTIPLICATION

Have students explore patterns with multiplication, rather than those based on addition. Give students a number of examples, e.g. 2, 4, 6, 8 and 2, 4, 8, 16, and have students identify the pattern and the factors. Ask students to explain how they worked out the pattern, e.g. compared two numbers and then tested with other sets of numbers in the pattern.

INDEPENDENT TASKS

Note: Choose from Tasks 1, 2 or 3.

You will need: poster paper, calculators, dice, long rolls of paper, Student Book p. 85 'Exploring Tables Patterns'

TASK 1: COMPARING PATTERNS

Have pairs of students develop a chart comparing patterns. Have them complete a pattern counting by 5s, then one multiplying by 5s. Have them write the differences in the numbers under the two patterns. Students repeat with at least three different sets of patterns. This activity could be extended by having students complete patterns with larger numbers. Note: students who require support may require the use of calculators.

TASK 2: INTERACTIVE TASK

Give pairs of students two dice and a long roll of paper. Have students roll the dice – one is the starting number and the other is the multiple. Students use these numbers to create a number pattern, continuing as far as they can go. Students may use calculators. Vary the activity by specifying a total amount of numbers for the pattern or a time limit.

TASK 3: STUDENT BOOK p. 85 *'Exploring Tables Patterns'*

TEACHING GROUP

You will need: BLM 1 'Hundred Chart', counters, NTO 4.1 'Hundred Chart', long rolls of paper, calculators

EXPLORING THE HUNDREDS GRID

- For students who require support, give them a copy of BLM 1 'Hundred Chart' and a number of counters. Have students explore different tables patterns by placing counters on the grid, and then creating patterns using multiplication. The patterns could be made on NTO 4.1 'Hundred Chart' and then screen shots taken and saved. Have students verbally describe the patterns. Make notes on the board identifying the differences.

PATTERNS WITH NUMBER LINES

- For students who require a challenge, give them long rolls of paper and have them create number lines showing the difference between addition patterns and multiplication patterns. Present the task in an open-ended format, allowing students to interpret. Students may need to use calculators.

REFLECTION

Select from the following to suit your class and their learning outcomes:

- Have students share their charts from Independent Tasks, Task 1, articulating the differences in their examples.
- Invite students to share their patterns from Independent Tasks, Task 2. Ask students to read their final numbers correctly and to explain the pattern. Ask, 'What was tricky at the end of the pattern? What did you notice between the numbers?'
- Have students share answers from Student Book p. 85 'Exploring Tables Patterns'. What did they discover? Different answers and responses could be examined using NTO 4.1 'Hundred Chart'.
- Invite students from the Teaching Group 'Patterns With Number Lines' to share their work. Ask, 'What is this number pattern? How did you work out what the next number would be?' Display these around the classroom.

LESSON PLAN 3

TUNING IN

EXPLORING OUR ENVIRONMENT

You will need: digital cameras or flip cameras

Take students outside the classroom to hunt for number patterns in their environment. This may include patterns on games, on clock faces, in art or classroom displays, at the canteen and so on. Have students take photos of the patterns. If they are using a flip camera, have them record the pattern while a partner comments on or describes it.

WHOLE-CLASS INTRODUCTION

SHARING OUR PATTERNS

You will need: an IWB or data projector, images/footage from Tuning In

Display the images/footage from Tuning In on the IWB or data projector. Have students describe why they took the image/footage and describe the pattern. Repeat with a number of examples.

INDEPENDENT TASKS

Note: Choose from Tasks 1, 2 or 3.

You will need: newspapers, magazines, poster paper, glue, scissors, *PowerPoint*, Student Book p. 86 'More Number Patterns'

TASK 1: IN THE PRESS

Have pairs of students look for number patterns in newspapers or magazines. Students cut out the images and create a collage on poster paper. Provide the extra challenge of finding patterns based on multiplication.

TASK 2: INTERACTIVE TASK

Have pairs of students use *PowerPoint* to present number patterns. On one slide, they present a pattern, then on the next slide, an explanation of the pattern. Set certain requirements for students, according to their abilities, e.g. create five different patterns containing at least six numbers – two patterns showing addition properties and three showing multiplication properties from certain starting points.

TASK 3: STUDENT BOOK p. 86 ***'More Number Patterns'***

TEACHING GROUP

You will need: counters, a digital camera, calculators

PRACTISING PATTERNS

- For students who require support, give them some counters and have them make number patterns based on addition, then multiplication. Once students have made the pattern with the counters, have them record it either by hand or with a digital camera. Then students write a description of the pattern, e.g. 'Counting by 2s starting at 8 and continuing for five numbers'. Repeat a number of times.

GOING BACKWARDS

- For students who require a challenge, give them larger starting numbers and have them explore patterns based on subtraction and division. Have students write the patterns and a description of them. Students may need to use calculators.

REFLECTION

Select from the following to suit your class and their learning outcomes:

- Have students present their collages from Independent Tasks, Task 1. Ask questions, such as 'Why did you select that particular image? What is the pattern in the sequence?'
- Have students share their *PowerPoint* presentations from Independent Tasks, Task 2. Students show their number pattern, then have the rest of the class determine the pattern before revealing the answer. Note: sometimes there is more than one answer.
- Invite students to share their counter patterns from the Teaching Group 'Practising Patterns', explaining how the pattern represents addition or multiplication.
- Invite students to share their backwards patterns from the Teaching Group 'Going Backwards' and what they learned.

Home Tasks

Select from the possible Home Tasks:

- Have students look for number patterns at home. Students note the patterns and bring them to school to share with the class.
- According to students' abilities, give them three starting numbers and counting sequences, and have them complete the patterns, and bring them back to school.

Assessment

- Have students complete **Student Assessment p. 87**.
- Review with students **Assessment Task Card 4.21**.

During the three lessons:

- Collect examples of students' hundreds grids or multiplication grids from Lesson Plan 1, Independent Tasks, Task 1, as evidence of identification of patterns, as well as evidence of the ability to work together.
- Collect examples of students' number patterns from Lesson Plan 2, Independent Tasks, Tasks 1 and 2, to illustrate their understanding of addition and multiplication patterns.
- Collect students' *PowerPoint* presentations from Lesson Plan 3, Independent Tasks, Task 2, as evidence of students' understanding of patterns.

Recommendations for Future Learning

Specific to Student Assessment p. 87; if the student is experiencing some difficulty:

Q 1 Review number patterns and how to work out the next number in the sequence. The student could be given a calculator to help.

Q 2 Review both forwards and backwards number patterns and how to determine the sequences.

Q 3 Have the student revisit how to describe number patterns using words. Remind the student to identify the starting number, the pattern and perhaps how many numbers there are.

Q 4 Review visual representations of number patterns, e.g. on Hundred Charts, multiplication charts or with counters. Have the student practise identifying the patterns. Resources such as NTO 4.1 'Hundred Chart' or BLM 30 'Multiplication Grid' could be used.

Q 5 Re-examine the collages from Lesson Plan 3 of everyday life or images from around the school.

If the student has not achieved the recommended skills for this unit:

1. See **Assessment Task Card 4.21** for specific recommendations.
2. Review basic number patterns, building to more complex and detailed number patterns.
3. Work with number patterns moving forwards, and then extend to those working backwards.
4. Support the student's development of patterns with the use of technology, e.g. calculators.
5. Review *Nelson Maths: Australian Curriculum NSW Year 3* Unit 21.

If the student has achieved the recommended skills and these skills are firmly established, consider:

1. Moving forward to *Nelson Maths: Australian Curriculum NSW Year 5* Unit 21.
2. Having the student work with larger numbers and more complex patterns.
3. Extending the student to work with decimals and fractions.

Unit 22 Angles

Measurement and Geometry
Angles MA2-16MG identifies, describes, compares and classifies angles

acute angle, angle, arm, obtuse angle, reflex angle, right angle, straight angle, vertex

LESSON PLAN 1

TUNING IN

INVESTIGATING RIGHT ANGLES

Give each student a sheet of paper. Have them move around the classroom and use the corner of the paper as a tester for right angles, e.g. corner of desk, corner of whiteboard. Have students record where they find the right angles on the sheet of paper.

WHOLE-CLASS INTRODUCTION

WHAT IS A RIGHT ANGLE?

You will need: sheet of paper from Tuning In, NTO 4.15 'Angles: Measuring'

Have students share where they found angles in the Tuning In activity. Examine any unusual ones. Ask, 'How did you check it was a right angle?' Lead into a more formal definition of a right angle. NTO 4.15 'Angles: Measuring' could be used to show the use of the protractor and value of 90° being a right angle.

INDEPENDENT TASKS

Note: Choose from Tasks 1, 2 or 3.

You will need: a digital camera, *Word*, Student Book p. 88 'Right Angles'

TASK 1: RIGHT ANGLES IN THE PLAYGROUND

Have pairs of students work in the playground, looking for and identifying right angles. Have them create a list of where the angles are found. Take photos of some of the angles identified by students.

TASK 2: INTERACTIVE TASK

Have pairs of students work on computers, using *Word* to create shapes and then identify and, if possible, label the right angles (if they exist in the shapes). This activity could be extended by having students use a Venn diagram to sort the shapes – those with only right angles and those without, and then those with both right angles and other angles.

TASK 3: STUDENT BOOK p. 88 *'Right Angles'*

TEACHING GROUP

You will need: LO: *L3535 'Ladybird mazes'*, LO: *L3505 'Turtle geometry'*

LADYBIRD MAZES

- For students who require support, have them use computers to work on LO: *L3535 'Ladybird mazes'*, manoeuvring a ladybird through a maze using forwards and backwards arrows and rotations of 90° and 45°.

TURTLE GEOMETRY

- For students who require a challenge, have them use computers to work on LO: *L3505 'Turtle geometry'*, using translations and rotations to make a turtle move and turn.

REFLECTION

Select from the following to suit your class and their learning outcomes:

- Display the photos of right angles found in Independent Tasks, Task 1, either electronically or on hard copy. Have students describe the common features of the different right angles.

- Referring to LO: *L3535 'Ladybird mazes'* or LO: *L3505 'Turtle geometry'*, discuss the relevance of a rotation of 90° to right angles. Ask, 'How is this different and how is it similar? Did you make shapes that contained right angles? How did you know it was a right angle?'
- Present a variety of angles on the board. Have students identify which of the angles are right angles, and why.

LESSON PLAN 2

TUNING IN

DIFFERENT ANGLES

Present a right angle and an acute angle on the board. Ask students to describe and compare the angles. Ask, 'What is the same and what is different?' Collect students' ideas on the board. Repeat with a right angle and a different acute angle, i.e. positioned differently or much smaller.

WHOLE-CLASS INTRODUCTION

TYPES OF ANGLES

You will need: NTO 4.15 'Angles: Measuring', display cards for each type of angle

Show NTO 4.15 'Angles: Measuring'. The arms on the angle can be moved to determine the range of the angle or you can allow the type of angle to be chosen randomly. Have students compare the angle with a right angle. Introduce the names of particular angles as they are revealed and put up the corresponding display card.

INDEPENDENT TASKS

Note: Choose from Tasks 1, 2 or 3.

You will need: poster paper, rulers, NTO 4.15 'Angles: Measuring', Student Book p. 89 'Types of Angles'

TASK 1: DRAWING ANGLES

Have pairs of students create a poster, using rulers to draw different types of angles (acute, right, straight, obtuse and reflex angles). Students place the angles randomly on the poster. Angles could be orientated in different directions. Have pairs swap their posters, and identify the types of angles on the other poster.

TASK 2: INTERACTIVE TASK

Have students use NTO 4.15 'Angles: Measuring' to practise identifying various angle types.

TASK 3: STUDENT BOOK p. 89 *'Types of Angles'*

TEACHING GROUP

You will need: wooden shapes (e.g. pattern blocks), BLM 59 'Angles in Triangles', protractors, calculators

EXPLORING ANGLES IN SHAPES

- For students who require support, give them a number of different wooden shapes so as to limit the types of angles to acute, right and obtuse angles. Have students identify the right angles first. When they feel comfortable with this, move onto acute and obtuse angles. Repeat with a number of shapes. Have students record this on a sheet of paper, by sketching or tracing the shapes.

ANGLES IN TRIANGLES

- For students who require a challenge, give them a copy of BLM 59 'Angles in Triangles'. Have them identify the angles and measure them with a protractor. Extend this to finding the sum of the angles in a triangle for those who are able to measure accurately.

REFLECTION

Select from the following to suit your class and their learning outcomes:

- Ask, 'Why do you think there are more right angles than, say, acute angles in our room?'
- Invite students to reflect on Independent Tasks, Task 1. Ask, 'When you drew a right angle, how did you make sure it was exact?'
- Have students share their work from the Teaching Group 'Exploring Angles in Shapes'. Ask, 'Why do you think you didn't find any straight or reflex angles?'
- Invite students from the Teaching Group 'Angles in Triangles' to share their findings.

TUNING IN

ANGLES IN THE ENVIRONMENT

You will need: digital cameras or flip cameras

Take students outside the classroom to hunt for angles in their environment. Set them a goal of finding a set number of right angles, angles greater than 90° and angles less than 90°. Have students record the angles they find either with a digital camera or flip camera.

WHOLE-CLASS INTRODUCTION

SHARING OUR ANGLES

You will need: an IWB or data projector, images/footage from Tuning In

Display the images/footage from Tuning In on the IWB or data projector. Have students describe why they took the image/footage and identify the angle they recorded, either as a right angle, an angle greater than 90° or an angle less than 90°. Repeat with a number of examples. The images/footage could be displayed on a digital photo frame. Introduce terms such as 'obtuse angle' and 'reflex angle' as descriptors of angles greater than 90°.

INDEPENDENT TASKS

Note: Choose from Tasks 1, 2 or 3.

You will need: magazines, newspapers, protractors, poster paper, scissors, glue, images/footage from Tuning In, Student Book p. 90 'Comparing Angles'

TASK 1: ANGLES IN THE PRESS

Have students look in magazines and newspapers for images that contain angles. Students may want to select an image and then identify and label all angles in the image, or they may want to find images that include specific angles, e.g. acute angles. Have students stick the image on poster paper, and label and describe the angles.

TASK 2: INTERACTIVE TASK

Using the digital pictures from Tuning In, have students use a protractor to measure the angles in their images. This could be done on a hard copy of the image. Alternatively, it is possible to add an image to an IWB and overlay a protractor. If the images have been taken on a tablet, there are protractor apps that allow students to measure the angle. Have students record the values with their images.

TASK 3: STUDENT BOOK p. 90 *'Comparing Angles'*

TEACHING GROUP

You will need: protractors, rulers, NTO 4.15 'Angles: Measuring'

DRAWING ANGLES

- For students who require support, give them a protractor, a sheet of paper and a ruler and have them practise drawing angles. Remind students to start with a base line. Show them how to read the scale on the protractor, make a mark and create the angle. NTO 4.15 'Angles: Measuring' could be used to aid explanations in the activity.

FINDING TOTALS OF ANGLES

- For students who require a challenge, have them draw a number of angles, e.g. 45°, 90°, 35°, 120°, then have them find the missing amount to make a straight line or an angle of 180°. Have students draw and label their diagrams (see example at right).

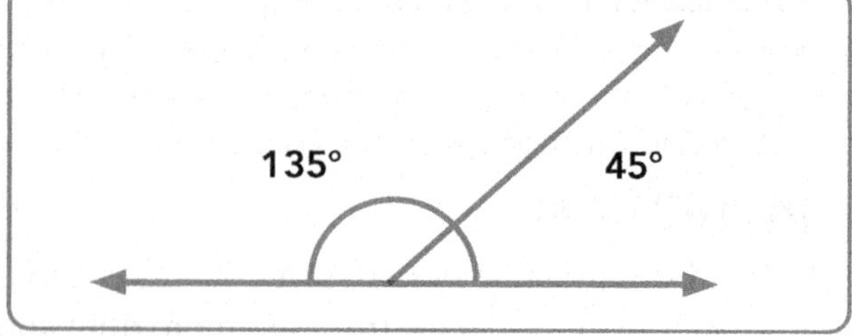

REFLECTION

Select from the following to suit your class and their learning outcomes:

- Have students share the images they found in Independent Tasks, Task 1.
- Have students share their analysis of the digital photos from Independent Tasks, Task 2.
- Invite students to share the angles they made that added to 180° in the Teaching Group 'Finding Totals of Angles'. Ask, 'How did you work out this value?'
- Draw a number of angles on the board and have students identify them as a right angle, an angle greater than 90° or an angle less than 90°. This activity could be extended by asking students to draw a particular angle, e.g. an obtuse angle, on the board.
- Use NTO 4.15 'Angles: Measuring' and revisit with students the use of a protractor when measuring angles.

Home Tasks

Select from the possible Home Tasks:

- Have students complete the table on BLM 60 'Drawing Angles', bring it to school and share it with the class.
- Have students complete a right-angle hunt at home with the corner of a sheet of paper. Have them record at least five items that have right angles.
- Have students complete BLM 61 'Timely Angles' and bring to school and review with the class.

Assessment

- Have students complete **Student Assessment p. 91**.
- Review with students **Assessment Task Card 4.22**.

During the three lessons:

- Collect examples of students' electronically generated materials, e.g. the investigation of shapes with right angles from Lesson Plan 1, Independent Tasks, Task 1. Add these to students' digital portfolios as evidence of their understanding of right angles.
- Collect students' photos/images of angles as evidence of their understanding of different types of angles.
- Make note of students who completed extension activities. Also make note of creative responses to different activities.
- Collect and review students' Home Tasks.

Recommendations for Future Learning

Specific to Student Assessment p. 91; if the student is experiencing some difficulty:

Q 1–2 Review what a right angle is. This could be via the activities or using protractors or NTO 4.15 'Angles: Measuring'. Examine angles that are greater than or less than a right angle and why. These comparisons could be made using protractors or estimation.

Q 3 Reinforce the identification of angle types by opening the cover of an old book to various angles and naming each angle as it appears.

If the student has not achieved the recommended skills for this unit:

1. See **Assessment Task Card 4.22** for specific recommendations.
2. Have the student practise identifying angles in shapes and objects as a measure of turn.
3. Have the student make visual comparisons between right angles and angles greater than or less than a right angle. The student could use tracing or baking paper to trace an angle and overlay on another to make the comparison.
4. Have the student work with protractors to measure values.
5. Review *Nelson Maths: Australian Curriculum NSW Year* 3 Unit 17.

If the student has achieved the recommended skills and these skills are firmly established, consider:

1. Moving forward to *Nelson Maths: Australian Curriculum NSW Year 5* Unit 22.
2. Having the student work with angles greater than 180°.
3. Extending the student to solve problems involving angles.

Unit 23 Equivalent Fractions

Number and Algebra

Fractions and decimals MA2-7NA represents, models and compares commonly used fractions and decimals

denominator, equivalent, fraction wall, fractions, numerator

LESSON PLAN 1

TUNING IN

HALF AND HALF AGAIN

You will need: coloured small kinder squares, scissors, glue, poster paper

Give students some small kinder squares. Have students fold and cut the squares in half. Ask, 'How do you know it is half?' Have them stick the squares on poster paper, and label each piece 'half'. Repeat, and continue to repeat to consolidate the concept of one half. Display the charts around the room.

WHOLE-CLASS INTRODUCTION

EQUIVALENT FRACTIONS

You will need: NTO 4.16 'Comparing Shapes'

On the board, draw two fraction representations, one of $\frac{1}{2}$ and the other of $\frac{2}{4}$. Ask, 'What is the same about these pictures? What is different?' Add a picture of $\frac{4}{8}$ and repeat the questions. Draw out the idea of equivalency. Repeat with a different shape, again highlighting the equivalency. Then use a non-equivalent example, e.g. $\frac{1}{2}$ and $\frac{3}{8}$. Ask, 'What is the same about these pictures? What is different? Collect students' ideas on the board. Introduce the term 'equivalence'.

INDEPENDENT TASKS

Note: Choose from Tasks 1, 2 or 3.

You will need: poster paper, coloured kinder squares, scissors, glue, *Word*, Student Book p. 92 'Halves, Quarters and Eighths'

TASK 1: DISPLAYING EQUIVALENT FRACTIONS

Give pairs of students poster paper and coloured kinder squares. Have them create a chart displaying equivalent fractions from the fraction family of halves, quarters and eighths. They could choose their own fractions, or students could be give fractions according to their abilities. Make sure the chart has a heading and that students label their fractions.

TASK 2: INTERACTIVE TASK

Have students work independently on computers, using the 'Shapes' function in *Word* to complete fraction representations of halves, quarters and eighths. Students could be allocated a fraction set, e.g. quarters, to investigate or they could look at the entire fraction family. Ensure students label their fractions.

TASK 3: STUDENT BOOK p. 92 *'Halves, Quarters and Eighths'*

TEACHING GROUP

You will need: fraction blocks, long strips of paper

FRACTION BLOCKS

- For students who require support, use fraction blocks to explore the concept of equivalence. Have students overlay blocks to form equivalent fractions. Then have students trace around the blocks to create diagrams with labels to record the equivalence. Focus on halves, quarters and eighths.

FRACTIONS ON A NUMBER LINE

- For students who require a challenge, have them create a number line, and place fractions – halves, quarters and eighths – on the number line. Have students consider and develop a solution for showing equivalent fractions, e.g. $\frac{1}{2}$ and $\frac{2}{4}$.

REFLECTION

Select from the following to suit your class and their learning outcomes:

- Have students share the fraction charts they made in Independent Tasks, Task 1. Ask, 'Are these two fractions equivalent? How do you know?' Look for different representations of ideas.
- Have students in the Teaching Group 'Fraction Blocks' share with the rest of the class, explaining what they did with the blocks and how they recorded the information.
- Invite students in the Teaching Group 'Fractions on a Number Line' to share their work with the group. Ask, 'How did you show equivalent fractions?' Have students explain ideas.

LESSON PLAN 2

TUNING IN

EXPLORING THIRDS AND SIXTHS

You will need: NTO 4.16 'Comparing Shapes'

Use NTO 4.16 'Comparing Shapes' to show and explore fractions of thirds and sixths. Enter a fraction for one of the shapes, and then have students enter an equivalent fraction. Repeat the process with students entering a non-equivalent fraction. Repeat with a number of examples.

WHOLE-CLASS INTRODUCTION

THIRDS AND SIXTHS

Brainstorm with students where you might see thirds and sixths in everyday life. This may include cutting an object or the fraction of an amount, e.g. $\frac{1}{3}$ of a cup in cooking, or a fraction of a packet of lollies. Gather ideas and record these on the board. This could be extended to indicate that thirds and sixths are a family of fractions, as are halves, quarters and eighths.

INDEPENDENT TASKS

Note: Choose from Tasks 1, 2 or 3.

You will need: strips of paper, scissors, split pins, LO: *L6542 'Exploring fractions'*, Student Book p. 93 'Fraction Strips'

TASK 1: FRACTION STRIPS

Give each student several strips of paper. Have students line the strips up and trim them to ensure they are all the same length. Have students divide one strip into thirds and one into sixths. (Note: students can be helped in this by having the strips of paper easily divisible by 3, e.g. 24 cm long.) Join these together with a split pin to form a family, with thirds on top and sixths underneath. Students should then be able to see the equivalent fractions. They may wish to colour the different segments. This activity could be extended to other fraction families, e.g. halves, quarters and eighths; fifths and tenths.

TASK 2: INTERACTIVE TASK

Have students work independently on computers, using LO: *L6542 'Exploring fractions'* where partially filled measuring cups are used to explore fractions: improper, mixed and equivalent.

TASK 3: STUDENT BOOK p. 93 *'Fraction Strips'*

TEACHING GROUP

You will need: strips of paper, split pins

THIRDS AND SIXTHS

- For students who require support, give them a sheet of paper and ask them to draw a shape and identify one third. Check that students are correct. If they are struggling, suggest they split a strip/block into thirds, as shown at right. Then ask them to identify one third, two thirds and three thirds. Have students draw an identical shape and divide into sixths.

EXTENDING THE FRACTION FAMILIES

- For students who require a challenge, have them work in pairs to extend the 'Fraction Strips' task to include twelfths and sixteenths. Have students add the new strips to their fraction families. To extend further, have students write equivalent fractions, including twelfths and sixteenths, using their strips.

REFLECTION

Select from the following to suit your class and their learning outcomes:

- Have students use their fraction strips from Independent Tasks, Task 1, to find fractions that are equivalent to ones called out.
- Have students share their fraction strips from the Teaching Group 'Thirds and Sixths', showing the fractions that are equivalent and explaining why.
- Write the term 'equivalent' on the board, and brainstorm with students what the term means. Have them include examples to reinforce points. NTO 4.16 'Comparing Shapes' could be used.

LESSON PLAN 3

TUNING IN

SHOWING EQUIVALENCE

You will need: A3 paper

Provide small groups or pairs of students with a fraction such as $\frac{5}{6}$ and a sheet of A3 paper.
Have students draw three different diagrams to show examples of equivalent fractions. Repeat with another fraction and have students draw the representations on the back of the sheet of paper.
Have students share with the rest of the class. This activity can be varied by providing different groups with different fractions.

WHOLE-CLASS INTRODUCTION

SORTING INTO GROUPS

You will need: BLM 39 'Fraction Cards'

Copy and cut up BLM 39 'Fraction Cards'. Give one card to each student. Have students organise themselves into similar fractions, e.g. thirds, then into fraction families. Finally, have them look for equivalent fractions.

INDEPENDENT TASKS

Note: Choose from Tasks 1, 2 or 3.

You will need: BLM 40 'Fraction Wall' enlarged to A3, fraction dice, counters, *Word*, BLM 41 'Build Your Own Fraction Wall', Student Book p. 94 'Painting a Fraction Wall'

TASK 1: THE FRACTION WALL

Give students an A3 copy of BLM 40 'Fraction Wall'. Have students colour and label the different sections of the wall. When finished, pairs of students could play a game with fraction dice, where they roll the dice and the number rolled represents the denominator, e.g. 4 represents $\frac{1}{4}$. Students then place a counter on the relevant fraction. The aim is to get one whole to win.

TASK 2: INTERACTIVE TASK

Have students work independently on computers, using the instructions on BLM 41 'Build Your Own Fraction Wall'.

TASK 3: STUDENT BOOK p. 94 *'Painting a Fraction Wall'*

TEACHING GROUP

You will need: BLM 40 'Fraction Wall' cut up as a jigsaw and laminated, whiteboard markers, strips of paper

BUILDING A FRACTION WALL

- For students who require support, copy and cut up BLM 40 'Fraction Wall' like a jigsaw (make this as complex as required for students' abilities). Laminate the pieces. Then have students put the wall back together. When complete, have them label each part with a whiteboard marker. Have a couple of versions, so students can repeat the activity.

EXTENDING TO TWELFTHS

- For students who require a challenge, have them consider how they could extend their fraction wall to twelfths. Give them strips of paper to complete and add to the bottom of their fraction wall.

REFLECTION

Select from the following to suit your class and their learning outcomes:

- Have students share their fraction walls from Independent Tasks, Task 1. Ask questions, such as, 'Find $\frac{2}{6}$ on your fraction wall. What is it equivalent to?'
- Mix up the jigsaws from the Teaching Group 'Building a Fraction Wall' and give one piece to each student. As a class, try to re-assemble the walls.
- Have students share the twelfths that they have added to their fraction walls in the Teaching Group 'Extending to Twelfths'. Invite them to explain how they worked out the divisions and where to place the strip.

Home Tasks

Select from the possible Home Tasks:

- Have students make their favourite sandwich and cut it into equivalent fractions. Then have them draw a picture of what they did.
- With the help of a parent or carer, have the student cut a piece of fruit into fractions, then write a description of what they can see.

Assessment

- Have students complete **Student Assessment p. 95**.
- Review with students **Assessment Task Card 4.23**.

During the three lessons:

- Make a copy of students' posters from Lesson Plan 1, Tuning In, noting the level of creativity.
- Collect copies of digital materials, e.g. the fraction wall produced in Lesson Plan 3, Independent Tasks, Task 2, to add to students' digital portfolios.
- Make a note of students' responses to the home tasks, and any difficulties they may have had.

Recommendations for Future Learning

Specific to Student Assessment p. 95; if the student is experiencing some difficulty:

Q 1 & 4 Review the activity of cutting the paper into different-looking halves. Have the student try to record this on paper. Note that all are showing half, so are equivalent.

Q 2–3 Review what 'equivalent' means. NTO 4.16 'Comparing Shapes' could be used here to support the student.

Q 5 Have the student spend more time working with the fraction wall. It may be beneficial to cut the wall into strips, and look at the construction as a series of strips stacked together. The student could sort the strips into fraction families to make the comparisons easier.

If the student has not achieved the recommended skills for this unit:

1. See **Assessment Task Card 4.23** for specific recommendations.
2. Review basic fraction concepts. Work with hands-on materials first, e.g. paper, fraction blocks and shapes, then move to pictorial representations and then numerical recording.
3. Use modelling equipment to reinforce ideas of equivalency, e.g. working with different divisions of the same shapes, so the shapes can be overlaid.
4. Review *Nelson Maths: Australian Curriculum NSW Year 3* Units 26 and 27.

If the student has achieved the recommended skills and these skills are firmly established, consider:

1. Moving forward to *Nelson Maths: Australian Curriculum NSW Year 5* Unit 23.
2. Extending the student in any of the listed activities by working with different fractions.
3. Having the student determine equivalent fractions without the aid of modelling materials or tools such as the fraction wall.

Unit 24 Counting with Fractions

Number and Algebra

Fractions and decimals MA2-7NA represents, models and compares commonly used fractions and decimals

equivalent, fractions, improper fractions, mixed numbers, number line, proper fractions

LESSON PLAN 1

TUNING IN

COUNTING BY FRACTIONS

You will need: long rolls of paper

Divide the class into groups. Give each group a fraction amount to count by, e.g. halves, quarters, thirds. Give the groups long rolls of paper, and have them start at zero and count by their designated fraction for as far as they can in the time allocated. Have groups share their counting strips.

WHOLE-CLASS INTRODUCTION

IMPROPER FRACTIONS TO MIXED NUMBERS

Write an improper fraction on the board. Ask students to identify the numerator and the denominator. Have students note that the numerator is larger than the denominator. Ask, 'Is this fraction simplified?' Work through with students how to convert the improper fraction to a mixed number. Name the mixed number and label it on the board. Repeat the conversion process a number of times, with students guiding the steps.

INDEPENDENT TASKS

Note: Choose from Tasks 1, 2 or 3.

You will need: long rolls of paper, *R10708 'Building Basic Skills'* (this can be found on the econtent repository http://econtent.thelearningfederation.com.au/ec/p/home), Student Book p. 96 'Improper Fractions to Mixed Numbers'

TASK 1: MIXED NUMBERS ON A NUMBER LINE

Give pairs of students long rolls of paper, a starting number, e.g. $1\frac{1}{2}$, and a number to count by, e.g. $\frac{1}{2}$. Have students continue the counting pattern. Then have them stop. Ask students to place these numbers on a number line created by them, with the divisions clearly marked. This activity can be varied according to students' abilities, by providing appropriate starting numbers and counting amounts.

TASK 2: INTERACTIVE TASK

Have students work independently on computers, using *R10708 'Building Basic Skills'*. This is a collection of 13 interactive learning objects, organised into four categories: introduction to fractions; naming fractions; fraction applications; and comparing and ordering fractions. Select the option that best caters for students' abilities, including working with mixed numbers if appropriate.

TASK 3: STUDENT BOOK p. 96 *'Improper Fractions to Mixed Numbers'*

TEACHING GROUP

You will need: BLM 42 'Fraction Cards: Mixed Numbers and Improper Fractions'

SORTING IMPROPER FRACTIONS AND MIXED NUMBERS

- For students who require support, copy BLM 42 'Fraction Cards: Mixed Numbers and Improper Fractions' and cut into cards. Have students sort the cards into piles of mixed numbers and improper fractions. Discuss what differentiates the groups. To extend the activity, students can convert the improper fractions into mixed numbers.

MORE TRICKY CONVERSIONS

- For students who require a challenge, give them more complex improper fractions to convert, e.g. $\frac{20}{9}$. Have students complete a number of conversions between improper fractions and mixed numbers. Have them compare answers to check each other's work.

REFLECTION

Select from the following to suit your class and their learning outcomes:

- Have students share the number lines they made in Independent Tasks, Task 1. Ask, 'How did you work out where the numbers went? Why did you write the numbers above/below the line?'
- Have students reflect on Independent Tasks, Task 2. Ask, 'What did you learn from completing the activity? What else would you like to learn about fractions?'
- Write a number of examples of improper fractions on the board. Have students identify them, then as a class have them simplify to mixed numbers. Select one fraction to use as the start of a counting sequence.

LESSON PLAN 

TUNING IN

SORTING CARDS

You will need: BLM 42 'Fraction Cards: Mixed Numbers and Improper Fractions', BLM 39 'Fraction Cards'

Copy and cut up BLM 42 'Fraction Cards: Mixed Numbers and Improper Fractions' and BLM 39 'Fraction Cards'. Give one card to each student and have them sort themselves into proper fractions, improper fractions and mixed numbers. Revise the terms if necessary. Collect the cards, redistribute and complete the sorting process again. This time, group them on the board.

WHOLE-CLASS INTRODUCTION

MIXED NUMBERS TO IMPROPER FRACTIONS

On the board, write several mixed numbers, with different but common denominators, e.g. halves, quarters and eighths. Have students identify them. Then say, 'Sometimes we need to work with mixed numbers as improper fractions. How can we convert this mixed number into an improper fraction?' Have students come up with suggestions. Then work through the process. Repeat with a number of examples.

INDEPENDENT TASKS

Note: Choose from Tasks 1, 2 or 3.

You will need: BLM 39 'Fraction Cards', BLM 42 'Fraction Cards: Mixed Numbers and Improper Fractions', BLM 43 'Symbols', *R10708 'Building Basic Skills'*, Student Book p. 97 'Mixed Numbers to Improper Fractions'

TASK 1: <, > OR =

Give groups of students cards from BLM 39 'Fraction Cards', BLM 42 'Fraction Cards: Mixed Numbers and Improper Fractions' and the symbols <, > and = from BLM 43 'Symbols'. Have students create statements with the cards, e.g. $\frac{1}{2} > \frac{1}{4}$. When the group agrees, have students record their statement on a sheet of paper. The activity could be further extended by having students find the cards that are less than 1 and those that are greater than $1\frac{1}{2}$. Note: review the symbols < and > if students are not familiar with them.

TASK 2: INTERACTIVE TASK

Have students work independently on computers, using *R10708 'Building Basic Skills'*. Select a different task to Lesson Plan 1, based on students' abilities. Note: this collection of 13 interactive learning objects is organised into four categories: introduction to fractions; naming fractions; fraction applications; and comparing and ordering fractions. Select the option that best caters for students' abilities.

TASK 3: STUDENT BOOK p. 97 *'Mixed Numbers to Improper Fractions'*

TEACHING GROUP

You will need: modelling equipment (e.g. Unifix blocks), LO: *L2w806 'Fraction fiddle: reach the target'*

MODELLING MIXED NUMBERS

- For students who require support, give them modelling equipment, e.g. Unifix blocks, and a mixed number such as $1\frac{1}{4}$. Have them represent the number, e.g. with small towers of Unifix. Repeat with a number of examples. Have them draw the models and write the accompanying mixed number.

LO: L2806 'FRACTION FIDDLE: REACH THE TARGET'

- For students who require a challenge, have them work independently on computers, using LO: *L2806 'Fraction fiddle: reach the target'*. Students need to help a boy to hit a bullseye with his paper plane by building up fractions. Note: there are various versions of this Learning Object to cater for different abilities.

REFLECTION

Select from the following to suit your class and their learning outcomes:

- Revisit the cards placed on the board in Tuning In. Select some improper fractions and have students convert them to mixed numbers. Then select some mixed numbers and have students convert them to improper fractions.
- Have students share their <, > and = statements from Independent Tasks, Task 1, with the group. Ask, 'How did you compare these two fractions? What did you do? How did you know this fraction was greater?'
- Have students from the Teaching Group 'Modelling Mixed Numbers' show their drawings to the group, or the models if they are still available.

LESSON PLAN **3**

TUNING IN

ORDERING FRACTIONS

You will need: BLM 39 'Fraction Cards', BLM 42 'Fraction Cards: Mixed Numbers and Improper Fractions'

Use the fraction cards from BLM 39 'Fraction Cards' and BLM 42 'Fraction Cards: Mixed Numbers and Improper Fractions'. Give each student a fraction card. Have students order themselves from smallest to largest. Check that they are correct. This activity could be made easier by taking out either the improper fractions or the mixed numbers.

WHOLE-CLASS INTRODUCTION

ORDERING FRACTIONS ON NUMBER LINES

You will need: BLM 39 'Fraction Cards', BLM 42 'Fraction Cards: Mixed Numbers and Improper Fractions'

Draw a number line on the board. Take some fraction cards (from BLM 39 'Fraction Cards' and BLM 42 'Fraction Cards: Mixed Numbers and Improper Fractions') and ask students to place them on the number line. Ask, 'Why did you place the card there? How do you know that card is larger/smaller?' Repeat a number of times. To challenge students, provide them with an empty number line to begin with.

INDEPENDENT TASKS

Note: Choose from Tasks 1, 2 or 3.

You will need: BLM 39 'Fraction Cards', BLM 42 'Fraction Cards: Mixed Numbers and Improper Fractions', long rolls of paper, scissors, glue, *PowerPoint*, Student Book p. 98 'Ordering Fractions'

TASK 1: LONG NUMBER LINES

Give groups of students copies of BLM 39 'Fraction Cards' and BLM 42 'Fraction Cards: Mixed Numbers and Improper Fractions'. Have them cut up the cards, order them and then place them on a large number line made from a long roll of paper. When they are happy, students can stick the cards down. This activity can be modified by removing certain cards, and it can be extended by having students add another five to ten fractions on their number lines.

TASK 2: INTERACTIVE TASK

Have students work independently on computers, using *PowerPoint* to create <, > and = fraction statements. Have students create the statement, and add an image that represents the fractions on the slide. Have students create five different slides with one statement on each. If students need ideas for fractions, BLM 39 'Fraction Cards' and BLM 42 'Fraction Cards: Mixed Numbers and Improper Fractions' can be used.

TASK 3: STUDENT BOOK p. 98 *'Ordering Fractions'*

TEACHING GROUP

You will need: BLM 39 'Fraction Cards', BLM 42 'Fraction Cards: Mixed Numbers and Improper Fractions'

ORDERING PRACTICE

- For students who require support, have them practise ordering cards. Use BLM 39 'Fraction Cards' and BLM 42 'Fraction Cards: Mixed Numbers and Improper Fractions', but start with three or four cards and slowly build up. Look at denominators and then numerators. Work with a small range, e.g. between zero and 1, before moving to mixed numbers.

WHERE IN REAL LIFE?

- For students who require a challenge, have them investigate how fractions are used in real life. Students could perform a search online or use media such as newspapers.

REFLECTION

Select from the following to suit your class and their learning outcomes:

- Have students share their long number lines from Independent Tasks, Task 1. Ask, 'Which were the trickiest fractions to place and why?'
- Have students share their *PowerPoint* presentations from Independent Tasks, Task 2.
- Invite students from the Teaching Group 'Where in Real Life?' to share what they found in their investigation.

Home Tasks

Select from the possible Home Tasks:

- Have students write and draw a description of a proper fraction, an improper fraction and a mixed number.
- Have students create their own fraction number line at home. Encourage them to show their parents or carers.

Assessment

- Have students complete **Student Assessment p. 99**.
- Review with students **Assessment Task Card 4.24**.

During the three lessons:

- Take photos of students' number lines or fraction sequences to include in their portfolios. Students may wish to include a comment with the photos about the work and their learning.
- Collect copies of digital materials, such as the *PowerPoint* statements in Independent Tasks, Task 2, to add to students' digital portfolios.
- Make a note of students' responses to Student Book p. 98 'Ordering Fractions', noting who can convert between improper fractions and mixed numbers, and who can order fractions correctly.

Recommendations for Future Learning

Specific to Student Assessment p. 99; if the student is experiencing some difficulty:

Q 1–3 Revisit the difference between mixed numbers and improper fractions. Have the student complete sorting and identification activities.

Q 4 Review how to convert from improper fractions to mixed numbers, by working with fractions such as $\frac{3}{2}$. Then build to more complex conversions. The student may need assistance with tables in the form of charts or grids, as the division relies on tables skills.

Q 5 Review how to convert from mixed numbers to improper fractions, by working with fractions such as $1\frac{1}{2}$. Then build to more complex conversions. The student may need assistance with tables in the form of charts or grids, as the multiplication relies on tables skills.

Q 6 Have the student practise identifying which fraction of a pair is smaller and larger, then extend to three fractions and so on. Initially, work with fractions that have common denominators, so numerators can be discussed, then work with fractions of families. The student could also model fractions to get a visual feel for the size of the fractions.

If the student has not achieved the recommended skills for this unit:

1. See **Assessment Task Card 4.24** for specific recommendations.
2. Review basic fraction concepts. Work with hands-on materials first, e.g. paper, fraction blocks and shapes, then move to pictorial representations and then numerical recording.
3. Use modelling equipment to reinforce smaller and larger fractions.
4. Have the student continue to practise ordering cards, beginning with familiar fractions with common denominators.
5. The student may need assistance with processes that rely on a knowledge of tables. Provide resources such as BLM 3 'Tables Chart 1' and BLM 4 'Tables Chart 2'.
6. Review *Nelson Maths: Australian Curriculum NSW Year 3* Units 26 and 27.

If the student has achieved the recommended skills and these skills are firmly established, consider:

1. Moving forward to *Nelson Maths: Australian Curriculum NSW Year 5* Unit 17.
2. Extending the student in any of the listed activities by working with more complex fractions.
3. Having the student create counting sequences and number lines working backwards.

Unit 25 Time

Measurement and Geometry

Time MA2-13MG reads and records time in one-minute intervals and converts between hours, minutes and seconds

am, analogue, days, digital, hours, measure, minutes, months, pm, seconds, time, weeks, years

LESSON PLAN 1

TUNING IN

CLOCKS

You will need: NTO 4.17 'Clocks: Advanced'

Use NTO 4.17 'Clocks: Advanced' to review the difference between analogue and digital time. Gather ideas and record these on the whiteboard, perhaps as a Venn diagram.

WHOLE-CLASS INTRODUCTION

ANALOGUE TIME

You will need: an analogue clock or NTO 4.17 'Clocks: Advanced'

Revise the different elements of an analogue clock, e.g. the hour and minute hands, the division between each number and what it represents, the number of minutes in an hour. Practise reading set times, including 'half past', 'quarter to' and 'quarter past'. Have students set times on the clock.

INDEPENDENT TASKS

Note: Choose from Tasks 1, 2 or 3.

You will need: BLM 44 'Clocks', internet access, scissors, Student Book p. 100 'Time'

TASK 1: INVESTIGATING TIME

Give each student a copy of BLM 44 'Clocks'. Give students some digital and analogue times and have them mark the times on the clocks. Alternatively, allow students to create their own, providing they include at least one 'half past', one 'quarter to' and one 'quarter past' time. When they have completed all of the clocks, have them cut out the cards, and stick them on a sheet of paper, ordering from earliest to latest.

TASK 2: INTERACTIVE TASK 

Have students work independently on computers to complete an activity on time in another language. These activities can be found by completing a search of 'time' and 'languages' on the econtent repository http://econtent.thelearningfederation.edu.au/ec/p/home. Select students' relevant LOTE language or one of interest.

TASK 3: STUDENT BOOK p. 100 *'Time'*

TEACHING GROUP

You will need: paper plates, scissors, cardboard for clock hands, split pins, NTO 4.17 'Clocks: Advanced'

ANALOGUE CLOCKS

- For students who require support, have them make an analogue clock with a paper plate. Discuss the positioning of the numbers, and the role of the hands. Talk about 'half past' and work through a number of examples. Repeat with 'quarter past' and 'quarter to'. NTO 4.17 'Clocks: Advanced' could also be used. Have students make the clocks, and then work in pairs to make times, using 'half past', 'quarter to' and 'quarter past'.

AT RIGHT ANGLES

- For students who require a challenge, pose the problem: 'What are the times when the hour hand and the minute hand are at right angles?' Students could use NTO 4.17 'Clocks: Advanced' either on the IWB as a group or individually on computers. Challenge students to look for the non-obvious answers as well.

REFLECTION

Select from the following to suit your class and their learning outcomes:

- Have students share their investigation from Independent Tasks, Task 1. Ask, 'Why did you select that time? How did you know to place this time after that one?'.
- Have students from the Teaching Group 'At Right Angles' share their results on the IWB. Ask, 'How do you know this is a right angle?'
- Display either analogue or digital times on NTO 4.17 'Clocks: Advanced', and invite students to the board to show the other time. Focus initially on 'half past', 'quarter past' and 'quarter to'.

LESSON PLAN 2

TUNING IN

ONE MINUTE

You will need: a class list, a stopwatch

Have all students stand up. Say, 'When you think one minute has passed, sit down.' Make sure there are no time devices in view. Set the time, and allow students to sit down at the appropriate time. At one minute, note on the class list the student/s who were the closest, but allow the activity to continue. After the activity has been completed, ask the student/s who were the closest if they had any strategies, and to share them with the class.

WHOLE-CLASS INTRODUCTION

TIME TERMS

You will need: NTO 4.3 'Calculator'

On the board, write the terms: 'second', 'minute', 'hour', 'day', 'week', 'month', 'year'. Brainstorm each of the terms and record ideas on the whiteboard. Discuss with students the relationship between each of the terms, e.g. 60 seconds = 1 minute, so in 3 minutes there are 180 seconds. NTO 4.3 'Calculator' could be useful for this activity – make sure the equation display is shown.

INDEPENDENT TASKS

Note: Choose from Tasks 1, 2 or 3.

You will need: poster paper, rulers, calculators, Student Book p. 101 'Converting Between Units of Time'

TASK 1: USING THE TIME

Have students work in pairs to make a poster listing all the time units, down one column of a table. In the subsequent columns, list things that are measured in the those time units. For example:

Time unit	Item 1	Item 2	Item 3
Seconds	Running 100m		
Minutes	Boiling an egg		
Hours	Flight between Melbourne and Perth		

Have students think of three examples for each time unit, including months and years. This activity could be extended to include composite examples, e.g. hours and minutes.

TASK 2: INTERACTIVE TASK 

Have students work independently on computers to create a conversion chart between each of the time units. For example: 1 minute = 60 seconds; 1 hour = 60 minutes. Allow students to use the program and format they prefer. They may need access to calculators.

TASK 3: STUDENT BOOK p. 101 *'Converting Between Units of Time'*

TEACHING GROUP

You will need: BLM 45 'Months', string, tape, calculators

ORDERING MONTHS

- For students who require support, make cards of the months from BLM 45 'Months'. Give the cards to students and have them place the months in order from January to December. On the back of each card, have students write something about the month. For example: the number of days in the month; the number of weeks in the month. Have students attach the cards to a piece of string to create a mobile.

COMPLETING THE TABLE

- For students who require a challenge, give them a calculator. Have them create the grid at right and complete for at least 10 days. To extend the activity, have students include 20 days, 50 days, 100 days and 1000 days.

Days	Hours	Minutes	Seconds
1	24	1440	
2	48		
3			

REFLECTION

Select from the following to suit your class and their learning outcomes:

- Have students share the posters they made in Independent Tasks, Task 1. Ask them questions to clarify their ideas: 'Why is boiling an egg measured in minutes? Could it be measured in seconds?' Look for unusual ideas, as well as common and recurring ones.
- Have students share the conversion charts they made in Independent Tasks, Task 2. Ask questions and have students use their conversion charts to find the answers.
- Repeat the activity from Tuning In. This time, have students sit with their eyes closed and put up their hand when they think one minute has passed. Then allow them to open their eyes. Did they improve?

LESSON PLAN 3

TUNING IN

WHAT'S THE TIME, MR WOLF?

Play a number of rounds of 'What's the Time, Mr Wolf?' One person (Mr Wolf) stands at the front, with their back to the group. The other students line up horizontally at the other end of the space, and ask, 'What's the time, Mr Wolf?' The person at the front calls out a time and the other students take that number of steps. This continues until they are close and then Mr Wolf yells out, 'Dinner time!' and chases students back to the starting point. If a student is tagged they become the new wolf. The activity could be extended using half and quarter times, with students taking half and quarter steps. Note: remind students not to be too rough in the chasing part.

WHOLE-CLASS INTRODUCTION

DIGITAL TIME: AM AND PM

Write 'am' and 'pm' on the board. Ask, 'What do these mean?' Gather ideas and record these on the board. Ask, 'Why do we use "am" and "pm"? Why is it important?'

INDEPENDENT TASKS

Note: Choose from Tasks 1, 2 or 3.

You will need: rulers, LO: *L9647 'Time tools: 24-hour to the minute: time match'*, Student Book p. 102 'am and pm Time'

TASK 1: MY TIMETABLE

Have students design and create a timetable of their day, from when they get up to when they go to bed. Have them use am and pm times. Encourage students to be creative. This could be done on the computer.

TASK 2: INTERACTIVE TASK

Have pairs of students work on computers, using LO: *L9647 'Time tools: 24-hour to the minute: time match'* to practise matching analogue and digital representations of 24-hour time. Note: there are various versions of this Learning Object to cater for different abilities.

TASK 3: STUDENT BOOK p. 102 *'am and pm Time'*

TEACHING GROUP

You will need: LO: *L9650 'Time tools: 12-hour to the minute: time match'*

LO: L9650 'TIME TOOLS: 12-HOUR TO THE MINUTE: TIME MATCH'

- For students who require support, have them work independently on computers, using LO: *L9650 'Time tools: 12-hour to the minute: time match'*. This is a modification of LO: *L9647 'Time tools: 24-hour to the minute: time match'*. There are also variations of LO: *L9650* activities to cater for different abilities.

INVESTIGATING PATTERNS

- For students who require a challenge, have them investigate times when the digits on a digital clock make a pattern, e.g. 12:34 or 13:57. Have students list the examples. Note: the times need to be realistic.

REFLECTION

Select from the following to suit your class and their learning outcomes:

- Have students share their timetables from Independent Tasks, Task 1. Clarify any ambiguities on the timetables. Check that times are realistic.
- Have students share their digital clock patterns from the Teaching Group 'Investigating Patterns'. Ask, 'Did you use a system to work out the patterns?'
- Play 'What's the Time, Mr Wolf?' again.

Home Tasks

Select from the possible Home Tasks:

- Have students create a timetable of one day from their weekend. Encourage them to be creative.
- Have students list all the time devices in their home, and where they are. For example: digital alarm clocks in the bedroom, timer on the stove in the kitchen, timer on the TV in the lounge room. To extend this activity, students could present their information in a table.

Assessment

- Have students complete **Student Assessment p. 103**.
- Review with students **Assessment Task Card 4.25**.

During the three lessons:

- Collect copies of the conversion charts from Lesson Plan 2, Independent Tasks, Task 2, to keep in students' digital portfolios.
- Collect samples of students' work, such as the tables of the time units from Lesson Plan 2, Independent Tasks, Task 1.
- Make note of any difficulties students had with the Student Book activities.

Recommendations for Future Learning

Specific to Student Assessment p. 103; if the student is experiencing some difficulty:

Q 1–2 Revisit the difference between digital and analogue time. Have the student practise linking the two using NTO 4.17 'Clocks: Advanced'. This could also be supported with paper clocks.

Q 3–4 Revisit links between each of the time values, such as 24 hours = 1 day. Student conversion charts could be revisited and further detail added.

Q 5 Revisit the difference between am and pm. Talk about what happens in the morning (am) and the afternoon/evening (pm). The student could benefit from matching activities using cards.

If the student has not achieved the recommended skills for this unit:

1. See **Assessment Task Card 4.25** for specific recommendations.
2. Work with hands-on materials, e.g. NTO 4.17 'Clocks: Advanced', paper and commercial clocks, to enable the student to become more familiar, particularly with the analogue variety.
3. Have the student create and then use matching activities, e.g. times in analogue, in digital and in words.
4. Put times in the context of the student's day, e.g. start school at 9 am, finish at 3:30 pm and so on.
5. Review *Nelson Maths: Australian Curriculum NSW Year 3* Unit 16.

If the student has achieved the recommended skills and these skills are firmly established, consider:

1. Having the student work with times to the minute in both digital and analogue format.
2. Moving forward to *Nelson Maths: Australian Curriculum NSW Year 5* Unit 25.
3. Having the student complete word problems involving time.

Unit 26 Time Problems

Measurement and Geometry

Time MA2-13MG reads and records time in one-minute intervals and converts between hours, minutes and seconds

am, analogue, arrival time, days, departure time, digital, hours, measure, minutes, months, pm, seconds, time, weeks, years

LESSON PLAN 1

TUNING IN

CLOCKS

You will need: BLM 46 'am and pm Times'

Copy and cut up the cards on BLM 46 'am and pm Times'. Give each student one card and have them order themselves from earliest in the day to latest in the day. Ask, 'What does "am/pm" mean? Why is it important to use "am" and "pm"?'

WHOLE-CLASS INTRODUCTION

SCHOOL TIMES

Brainstorm some of the important times in a school day, e.g. the start of the day, the start of lunch, literacy time. Gather and record ideas and times on the board. Ask, 'How could we organise these times?' (They could be organised on a timeline or in a timetable.) Highlight the importance of 'am' and 'pm'.

INDEPENDENT TASKS

Note: Choose from Tasks 1, 2 or 3.

You will need: Student Book p. 104 'Ashleigh's Day'

TASK 1: LENGTH OF SCHOOL ACTIVITIES

Give students a list of school activities, and ask them to calculate how long each one takes. For example: the length of school day; the length of recess and lunchtime; the amount of time spent on maths; the amount of time spent on literacy; the amount of time spent in specialist classes. Review the difference between minutes and hours.

TASK 2: INTERACTIVE TASK

Have students work independently on computers to create a timetable for the day, including home activities, school activities and after-school activities. Have students identify the am and pm times. Students could be guided by a template, or you could give them flexibility in the creation and design.

TASK 3: STUDENT BOOK p. 104 *'Ashleigh's Day'*

TEACHING GROUP

You will need: a TV guide or a transport timetable, highlighter pens

READING AM AND PM TIME

- For students who require support, give them a schedule such as a TV guide or a transport timetable. Ask them to find and locate things on the timetable using am and pm time. When they have located each item, have students highlight it. Repeat a number of times, until students are confident with the times.

SCHOOL ACTIVITIES FOR A WEEK

- For students who require a challenge, have them work out how much time is spent in a week on various school activities, e.g. at school, in breaks (recess, lunchtime), learning, in specialist sessions, at sport. Have students show their working.

REFLECTION

Select from the following to suit your class and their learning outcomes:

- Review students' work on the length of school activities in Independent Tasks, Task 1. This could be extended to include the work students completed in the Teaching Group 'School Activities for a Week'. Ask, 'How did you find that amount? Was all that time in the morning or afternoon? Why are "am" and "pm" important?'
- Work through the answers for Student Book p. 104 'Ashleigh's Day'. Have students share the additional activities they came up with.
- Have students share their timetables from Independent Tasks, Task 2, either electronically or on hard copy. Ask questions about the time divisions, and the importance of 'am' and 'pm'.

LESSON PLAN 2

TUNING IN

ONE MINUTE

You will need: sticky notes

Give each student a sticky note and have them write on it how long it takes them to get to school, in minutes. Have students order themselves from the shortest to the longest time.

WHOLE-CLASS INTRODUCTION

FROM HERE TO THERE

Ask, 'How do we measure time between two locations?' Give examples, e.g. from home to school; from home to the nearest city; from Sydney to Perth. Gather ideas and record these on the board. Look at what factors are important, e.g. knowing the starting time and the finishing time, and working out the difference; knowing the units of time measured in; knowing if the times are am or pm.

INDEPENDENT TASKS

Note: Choose from Tasks 1, 2 or 3.

You will need: an interactive map tool, Student Book p. 105 'Adelaide to Darwin'

TASK 1: TIMING OF A GAME

Pose the following problem to students: 'A football match takes 2 hours and 15 minutes. What might be suitable starting times and finishing times?' Have pairs of students create a chart, showing their ideas, including how the game is broken up. Have them give reasons for the starting and finishing times.

TASK 2: INTERACTIVE TASK

Using an interactive map tool, have students investigate how long it takes to travel between two locations. If students key in the locations, then select the 'Get directions' button, suggested routes and travel times will come up. Students could be provided with locations to investigate or they could investigate their own. Have students note the locations, the distance between them and the time to travel that distance. This activity could be extended by having students collect the data in a table.

TASK 3: STUDENT BOOK p. 105 *'Adelaide to Darwin'*

TEACHING GROUP

You will need: an interactive map tool, BLM 12 'Map of Australia'

LOCAL MAPS AND TIMES

- For students who require support, use a local map (which could be found online) and allocate a starting time, e.g. 9 am, and a location, e.g. a local shop. Provide a finishing location, e.g. school, and a finishing time, e.g. 9:17 am. Have students calculate the time between the two locations. Have them complete this for a number of locations, ensuring they record the starting time, finishing time and travel time, as well as the two locations. This activity could be extended by having students complete a table to collect the data.

PLANNING A HOLIDAY

- For students who require a challenge, have them plan a driving holiday around Australia. Students plan the route and estimate the travel time between each location, using an interactive map tool. Students could record the information on BLM 12 'Map of Australia'.

REFLECTION

Select from the following to suit your class and their learning outcomes:

- Have students share their work from Independent Tasks, Task 1. Ask, 'Why did you decide to start then? Is that a reasonable finishing time?'
- Have students share their interactive map tool investigations from Independent Tasks, Task 2. If students have used specified locations, compare answers. If students have chosen their own locations, examine the locations and the times between.
- Ask questions and record ideas on the board about why it is important to know the travel time as well as the distance. Ask, 'Do you think we could work out the travel time if we know the distance?' (Relate this to speed.) 'What other factors are important?'

LESSON PLAN 3

TUNING IN

ORDERING 24-HOUR TIME

You will need: BLM 47 '24-Hour Times'

Ask, 'What is 24-hour time?' Gather ideas and record these on the board and show students how 24-hour time is organised. Copy and cut up BLM 47 '24-Hour Times'. Give one card to each student. Have students read the times on the cards, and then order themselves from 0100 to 2400. Help students who are experiencing difficulty with working out the times.

WHOLE-CLASS INTRODUCTION

TIMETABLE

You will need: BLM 48 'Train Timetable'

Give students a copy of BLM 48 'Train Timetable'. Have students practise reading the times from the timetable. Ask questions, such as, 'What time does the first train leave South Station? What time does the last train arrive at North Station?' Discuss how the times are presented (in 24-hour time). Use today's date as the date for the journey. Investigate various ways of writing the date correctly in short and long forms such as 7 August 2015, and 07.08.2015.

INDEPENDENT TASKS

Note: Choose from Tasks 1, 2 or 3.

You will need: BLM 49 'Plane Flights', LO: *L9892 'Timetable: talent quest: travel'*, Student Book p. 106 'Television Time'

TASK 1: PLANE FLIGHTS

Give each student a copy of BLM 49 'Plane Flights'. Have students plan a trip based on a scenario such as: 'I need to fly to Melbourne to attend a wedding. The wedding is at 6 pm, and I need to be in Melbourne at least two hours before that, so I can get from the airport to the hotel and then to the wedding. I need to be back in Brisbane the following day to start work at 8 pm.' Have students work out which flights to catch and why, how long the flights take and the total cost.

TASK 2: INTERACTIVE TASK

Have students work individually on computers, using LO: *L9892 'Timetable: talent quest: travel'* to help plan a talent quest. Students check the travel plans for each act and work out their earliest arrival times. Note: there are various versions of this Learning Object to cater for different abilities.

TASK 3: STUDENT BOOK p. 106 *'Television Time'*

TEACHING GROUP

You will need: BLM 48 'Train Timetable' or BLM 49 'Plane Flights' or a local transport timetable, LO: *L7900 'Timetable: extreme talent quest: travel'*

READING TIMETABLES

- For students who require support, give them copies of BLM 48 'Train Timetable' or BLM 49 'Plane Flights' or a local transport timetable. Have them practise reading the timetables, guiding them to read column and row headings. Talk about the representation of the times, and the duration of the trips. Set some questions about the particular timetable, then have students go through the answers as a group.

LO: L7900 'TIMETABLE: EXTREME TALENT QUEST: TRAVEL'

- For students who require a challenge, have them work individually on computers, using LO: *L7900 'Timetable: extreme talent quest: travel'* to help plan an extreme talent quest. Students check the travel plans for each act and work out their earliest arrival times.

REFLECTION

Select from the following to suit your class and their learning outcomes:

- Have students share their flight details from Independent Tasks, Task 1, justifying the times and the cost. Check to see that students have worked out the length of each flight.
- Have students reflect on the Learning Object activities. Ask, 'What was tricky? How was time important? How did you work out the answers?'
- Have students plan a local trip, perhaps associated with an excursion. Ask, 'How long will the trip take? How could we work it out? Could we take public transport? Do you think it would be cheaper or more expensive to take public transport?'

Home Tasks

Select from the possible Home Tasks:

- Have students determine how much time they spend sleeping over five days. Have them record the time for each day, and then find the total.
- Have students make a list of all the TV shows they watch for a week. Then have them use a TV guide to work out the length of time they have spent watching TV.

Assessment

- Have students complete **Student Assessment p. 107**.
- Review with students **Assessment Task Card 4.26**.

During the three lessons:

- Collect copies of students' timetables from Lesson Plan 1, Independent Tasks, Task 1, for their portfolios.
- Collect samples of students' extension or support materials, e.g. from the Lesson Plan 2 Teaching Group 'Planning a Holiday', for their portfolios.
- Collect students' responses to the Lesson Plan 3 Reflection task of planning a local trip, as evidence of their understanding of the content from the unit.

Recommendations for Future Learning

Specific to Student Assessment p. 107; if the student is experiencing some difficulty:

Q 1 Review the different time formats: analogue, digital, 24-hour, am and pm. Using cards created from BLM 44 'Clocks', BLM 46 'am and pm Times' and BLM 47 '24-Hour Times', have the student complete sorting and ordering activities to increase familiarity with the time representations.

Q 2 Revisit the concept of a daily timetable showing what activities occur and when. The student could be given a grid with different times of the day, and draw or find a picture to represent a suitable activity.

Q 3–4 Review how to read timetables. Have the student look at the time format, how the information is organised and presented and how to locate information. Practise with a variety of different TV schedules. Review the correct way to write the date.

If the student has not achieved the recommended skills for this unit:

1. See **Assessment Task Card 4.26** for specific recommendations.
2. Work with a variety of different timetables and schedules. Begin with one the student is familiar with, extending to ones that are unfamiliar.
3. Put times in the context of the student's day, e.g. start school at 9 am, finish at 3:30 pm and so on.
4. Continue practising with different time formats and representations so the student becomes familiar with all forms. Matching activities could be developed using BLM 44 'Clocks', BLM 46 'am and pm Times' and BLM 47 '24-Hour Times'.
5. Review *Nelson Maths: Australian Curriculum NSW Year 3* Unit 16.

If the student has achieved the recommended skills and these skills are firmly established, consider:

1. Having the student investigate how timetables and schedules are represented in different countries, through internet research.
2. Moving forward to *Nelson Maths: Australian Curriculum NSW Year 5* Unit 26.
3. Having the student complete problems that involve multiple steps.

Unit 27 Fractions and Decimals

Number and Algebra
Fractions and decimals MA2-7NA represents, models and compares commonly used fractions and decimals

decimal, decimal point, fractions, hundredths, larger, number line, order, place value, smaller, tenths, units, whole number, zero

LESSON PLAN 1

TUNING IN

SORTING DECIMALS

You will need: BLM 34 'Decimal Numbers and Words 1', BLM 35 'Decimal Numbers and Words and 2', scissors

Give each student a copy of BLM 34 'Decimal Numbers and Words 1'and BLM 35 'Decimal Numbers and Words and 2'. Have them cut up the cards and sort them into two groups – numbers that just have tenths and numbers that have both tenths and hundredths. Check sorting with students.

WHOLE-CLASS INTRODUCTION

WHAT IS A TENTH?

You will need: Unifix blocks

Ask, 'What is a tenth?' Have students brainstorm ideas and record these on the board. Draw out the representations, e.g. one tenth, $\frac{1}{10}$ and 0.1. Show students some Unifix blocks. Ask, 'How could we show one tenth?' (10 Unifix blocks in a column and one in a different colour) Now ask, 'What is two tenths?' and so on. Extend the activity by looking at $3\frac{7}{10}$.

INDEPENDENT TASKS

Note: Choose from Tasks 1, 2 or 3.

You will need: poster paper, rulers, LO: *L1085 'Hopper: tenths'*, Student Book p.108 'Tenths'

TASK 1: LISTING TENTHS

Give students poster paper and rulers. Have them create three columns and label them as shown at right. Have students continue to ten tenths (one whole) and then continue with $1\frac{1}{10}$ and so on. Have students extend the chart as far as they can. To extend the activity, students could select three examples and draw pictures to illustrate the three numbers.

In Words	As a Fraction	As a Decimal
one tenth	$\frac{1}{10}$	0.1

TASK 2: INTERACTIVE TASK

Have students work independently on computers, using LO: *L1085 'Hopper: tenths'* to help a frog to jump along a number line using patterns of tenths.

TASK 3: STUDENT BOOK p. 108 *'Tenths'*

TEACHING GROUP

You will need: Unifix blocks, BLM 34 'Decimal Numbers and Words 1', BLM 35 'Decimal Numbers and Words and 2', LO: *L868 'Wishball: Tenths'*

MODELLING TENTHS

- For students who require support, provide Unifix blocks. Give students an amount, e.g. $\frac{3}{10}$, and have them model this with the blocks. Repeat a number of times. Then give them a card with a number that has tenths (no hundredths) and have them model this number. BLM 34 'Decimal Numbers and Words 1' and BLM 35 'Decimal Numbers and Words 2' could be used for this.

LO: L868 'WISHBALL: TENTHS'

- For students who require a challenge, have them work in pairs on computers, using LO: *L868 'Wishball: tenths'*. Students receive a starting number and work towards turning it into a target number.

REFLECTION

Select from the following to suit your class and their learning outcomes:

- Have students present their posters from Independent Tasks, Task 1. Ask questions where students can refer to their charts, e.g. 'Who can write the fraction that is the same as five tenths?'
- Invite students to share their modelled tenths from the Teaching Group 'Modelling Tenths' and explain their representations. Encourage them to use the correct terminology.
- Write a number of decimals with tenths and have students write the corresponding fractions or decimals in words. Extend by including mixed numbers.

LESSON PLAN 

TUNING IN

SMALLER OR LARGER?

You will need: Unifix blocks

Write a pair of fractions on the board, e.g. $\frac{1}{10}$ and $\frac{1}{100}$. Ask, 'Which one is larger?' Explore a couple of pairs, e.g. $\frac{2}{10}$ and $\frac{20}{100}$, $\frac{7}{10}$ and $\frac{7}{100}$. Repeat with 0.1 and 0.01, 0.2 and 0.002. Expand to include words as well, e.g. one tenth and two tenths. Support students with visuals such as Unifix blocks or diagrams.

WHOLE-CLASS INTRODUCTION

WHAT IS A HUNDREDTH?

You will need: NTO 4.14 'Hundred Chart: FDP'

Ask, 'What is a hundredth?' Have students brainstorm ideas and record these on the board. Draw out the representations, e.g. one hundredth, $\frac{1}{100}$ and 0.01. Show students NTO 4.14 'Hundred Chart: FDP' and have them shade the relevant fraction or decimal and vice-versa.

INDEPENDENT TASKS

Note: Choose from Tasks 1, 2 or 3.

You will need: BLM 50 'Hundreds Squares', LO: *L587 'Hopper: hundredths'*, Student Book p. 109 'Hundredths'

TASK 1: HUNDREDS SQUARES

Give each student a copy of BLM 50 'Hundreds Squares' and six decimals or fractions. Have students note the decimal or fraction, shade the grid and state the missing representation. Extend by giving students a mixed number.

TASK 2: INTERACTIVE TASK

Have students work independently on computers, using LO: *L587 'Hopper: hundredths'* to help a frog to jump along a number line using patterns of hundredths.

TASK 3: STUDENT BOOK p. 109 *'Hundredths'*

TEACHING GROUP

You will need: BLM 50 'Hundreds Squares', LO: *L1090 'Hopper challenge: ultimate'*

MODELLING MIXED NUMBERS

- For students who require support, give them each a copy of BLM 50 'Hundreds Squares'. Give each student different fractions, e.g. $\frac{39}{100}$, and have them colour the squares and complete the labels. Check students are colouring correctly. Then provide another copy of BLM 50 'Hundreds Squares' and have them shade each square and then swap with a partner, who has to complete the labels. If students are struggling, work with the fraction representation first, then make links to the decimals.

LO: L1090 'HOPPER CHALLENGE: ULTIMATE'

- For students who require a challenge, have them work independently on computers, using LO: *L1090 'Hopper challenge: ultimate'* to help a frog to jump along a number line using patterns of either whole numbers, tenths or hundredths.

REFLECTION

Select from the following to suit your class and their learning outcomes:

- Have students share their hundreds squares from Independent Tasks, Task 1. Ask questions, such as, 'Do you need to start colouring from the same starting point?' (No.) Check that students have the correct decimal and fraction equivalent. Students who completed the hundreds squares in the Teaching Group 'Modelling Mixed Numbers' may also like to share.
- Have students share their experiences with the Hopper activities. Ask, 'What was difficult about the activity? Was your estimate correct? Why?'
- Write a number of decimals with hundredths and have students write the corresponding fractions or decimals in words. Extend students by including mixed numbers.

LESSON PLAN 3

TUNING IN

CREATING DECIMALS

You will need: three 10-sided dice for each pair, six sheets of card or paper, a number line numbered from 0 to 10 drawn on the board

Give pairs of students three 10-sided dice. Students roll the dice to create and record numbers with two decimal places, e.g. 4.36. Have them collect six different numbers, then order the numbers from smallest to largest. Have students share their answers. Check the order. Choose six of the students' numbers to write onto cards and have students place the numbers at the approximately correct place on a number line.

WHOLE-CLASS INTRODUCTION

CHANGING TO FRACTIONS

You will need: the six numbers from Tuning In, NTO 4.14 'Hundred Chart: FDP'

Using the six numbers from Tuning In, have students find the equivalent fractions. Use NTO 4.14 'Hundred Chart: FDP' to support students with the conversions. The task could be extended by having students write their numbers in words. Have students share their answers and articulate the conversion process.

INDEPENDENT TASKS

Note: Choose from Tasks 1, 2 or 3.

You will need: BLM 50 'Hundreds Squares', modelling equipment (e.g. Unifix blocks, MAB), a digital camera, LO: *L7901 'Swamp survival: hundredths counting'*, Student Book p. 110 'Decimals and Fractions'

TASK 1: MODELLING DECIMALS

Give students modelling equipment, e.g. Unifix blocks and MAB, as well as BLM 50 'Hundreds Squares'. Provide each student with six different decimals, some with tenths and some including hundredths, and have students model their numbers with the equipment. If suitable, include mixed numbers. Allow students to select the equipment they wish to use. Have students either draw their model and write the decimal or have them take a photo (don't forget to include the student's name in the photo for identification purposes later). Have students then convert their decimal to a fraction, with the support of BLM 50 'Hundreds Squares' if required. Have students retain at least one of their models for sharing purposes.

TASK 2: INTERACTIVE TASK

Have students work independently on computers, using LO: *L7901 'Swamp survival: hundredths counting'* to order decimals so the boy can cross the swamp. Note: there are various versions of this Learning Object to cater for different abilities.

TASK 3: STUDENT BOOK p. 110 *'Decimals and Fractions'*

TEACHING GROUP

You will need: NTO 4.14 'Hundred Chart: FDP', LO: *L872 'Wishball challenge: tenths'* or LO: *L873 'Wishball challenge: hundredths'*

HUNDREDS SQUARE PRACTICE

- For students who require support, show them a shaded grid on NTO 4.14 'Hundred Chart: FDP'. Have them identify the fraction, then state the fraction in words, and then state the decimal. Repeat a number of times. Then give students a fraction and have them model it on the IWB. Repeat a number of times.

WISHBALL CHALLENGE

- For students who require a challenge, have them attempt LO: *L872 'Wishball challenge: tenths'* or LO: *L873 'Wishball challenge: hundredths'*, according to their ability.

REFLECTION

Select from the following to suit your class and their learning outcomes:

- Have students share one of their decimal models from Independent Tasks, Task 1, explaining why they selected the equipment and what the model shows.
- Spend time reviewing the answers to Student Book p. 110 'Decimals and Fractions'. Check any problem areas with students and revisit on the board.
- Display a fraction on NTO 4.14 'Hundred Chart: FDP' and have students state the related decimal. Repeat a number of times. Then state a decimal and have students show the related fraction.

Home Tasks

Select from the possible Home Tasks:

- Have students look for decimals to two decimal places in a newspaper (hard copy or online). Have them record the decimals and convert them to fractions.
- Give students a copy of BLM 35 'Decimal Numbers and Words 2'. Have students cut out the cards and order from smallest to largest and write the fractions underneath (with the help of parents or carers).

Assessment

- Have students complete **Student Assessment p. 111**.
- Review with students **Assessment Task Card 4.27**.

During the three lessons:

- Keep a copy of samples, such as the tenths chart from Lesson Plan 1, Independent Tasks, Task 1, for student portfolios.
- Collect copies of digital materials, such as the photos of modelled decimals from Lesson Plan 3, Independent Tasks, Task 2, to add to students' digital portfolios.
- Have students list three things they can do in the topic 'Decimals and Fractions' and three things they would like to learn more about.

Recommendations for Future Learning

Specific to Student Assessment p. 111; if the student is experiencing some difficulty:

Q 1 Using NTO 4.14 'Hundred Chart: FDP', have the student practise representing fractions and decimals, and then identifying the coloured squares as fractions and decimals. This could be extended by having the student colour or record the fractions and decimals on BLM 50 'Hundreds Squares'.

Q 2 Have the student revisit representing fractions, then decimals, with modelling equipment, e.g. Unifix blocks and MAB. Discuss different ways of recording this.

Q 3–4 Have the student practise completing matching activities, perhaps with cards. These could be generated using BLM 34 'Decimal Numbers and Words 1' and BLM 35 'Decimal Numbers and Words 2' and adding in the fraction equivalents.

Q 5 Practise with decimals in which there are no hundredths.

If the student has not achieved the recommended skills for this unit:

1. See **Assessment Task Card 4.27** for specific recommendations.
2. Review basic place-value concepts, revisiting tenths then hundredths.
3. Work with only tenths, converting between decimals and fractions until ideas are consolidated.
4. Have the student continue to work with modelling equipment and hundreds squares. Hundreds squares could be incorporated into matching activities.
5. Use calculators to support the student with the conversions from fractions to decimals.
6. Review *Nelson Maths: Australian Curriculum NSW Year 3* Units 26, 27 and 28.

If the student has achieved the recommended skills and these skills are firmly established, consider:

1. Moving forward to *Nelson Maths: Australian Curriculum NSW Year 5* Unit 27.
2. Extending the student in any of the listed activities by working with mixed numbers.
3. Extending the student into thinking about thousandths.

Unit 28 Number Sentences

Number and Algebra

Patterns and algebra MA2-8NA generalises properties of odd and even numbers, generates number patterns, and completes simple number sentences by calculating missing values

addition, number sentences, subtraction, unknown quantity

LESSON PLAN 1

TUNING IN

REVIEWING ADDITION

On the board, write some addition equations of varying difficulty, according to students' abilities. For example: 12 + 5 =; 21 + 38 =; 47 + 59 =; 125 + 287 =; What is 28 plus 49?; Tony had 125 toy cars and 158 figurines. How many items did he have altogether? Have students record the equations and work out the answers. Discuss the different strategies they used to solve the equations.

WHOLE-CLASS INTRODUCTION

MISSING NUMBERS – ADDITION

You will need: sticky notes

Write the equation 20 + 10 = 30 on the board. Cover the 10 with a sticky note. Then write the equation 20 + ☐ = 30 underneath. Ask, 'What is the missing number? How could we work this out?' Discuss strategies with students and record them on the board. Repeat with another equation, e.g. 15 + ☐ = 21. Repeat with a number of examples.

INDEPENDENT TASKS

Note: Choose from Tasks 1, 2 or 3.

You will need: BLM 51 'Number Cards', LO: *L9927 'Thinking addition: 2-digit plus 1-digit: assessment'*, LO: *L9926 'Thinking addition: 2-digit plus 2-digit: assessment'*, Student Book p. 112 'Missing Numbers – Addition'

TASK 1: CARD EQUATIONS

Give pairs or small groups of students a set of number cards from 1 to 50 and a target number. These could be made from BLM 51 'Number Cards'. Students mix up the cards, then one of them picks the top card. If 38 is picked, the student writes 38 + ☐ = [target number, e.g. 60]. Students then need to solve the equation. Once the answer is found, students continue taking turns. This activity could be varied by reducing the number of cards or changing the target number.

TASK 2: INTERACTIVE TASK

Have students work independently on computers, using LO: *L9927 'Thinking addition: 2-digit plus 1-digit: assessment'* and then LO: *L9926 'Thinking addition: 2-digit plus 2-digit assessment'* to test their ability to turn word problems into equations.

TASK 3: STUDENT BOOK p. 112 *'Missing Numbers – Addition'*

TEACHING GROUP

You will need: sticky notes, BLM 51 'Number Cards'

BASIC FACTS

- For students who require support, work with basic fact equations. Have students write each element of an equation on a sticky note, e.g. 3 + 5 = 8. Then have them turn the 5 over. Have them record the new equation 3 + ☐ = 8. Ask, 'What is the ☐?' Discuss. Then repeat, turning the 3 over. Repeat with some more equations, working on basic addition facts and missing numbers in the equations.

CARD PLUS EQUATIONS

- For students who require a challenge, use number cards from BLM 51 'Number Cards'. Give them a target number. Have students draw out two numbers from the number cards, e.g. 35 and 12, and create a three-part equation: 35 + 12 + □ = [target number, e.g. 120]. Then have students solve the equation. Repeat a number of times, then have students construct the equation as 35 + □ + 12 = 120 and solve. Ask, 'Does the order of the numbers make any difference?'

REFLECTION

Select from the following to suit your class and their learning outcomes:

- Have students reflect on the card activity from Independent Tasks, Task 1, or from the Teaching Group 'Card Plus Equations'. Say, 'Explain how you found the answer for this equation. What strategies did you use?' Keep an eye out for errors and review with students as required.
- Review the answers to Student Book p. 112 'Missing Numbers – Addition'. Have students share strategies for finding missing numbers, particularly for Questions 3 and 5.
- Write a couple of examples on the board, e.g. 24 + □ = 50 and 60 = 21 + 15 + □. Have students discuss how they would find the missing numbers. Stress the importance of checking answers.

LESSON PLAN 2

TUNING IN

REVIEWING SUBTRACTION

On the board, write some subtraction equations of varying difficulty, according to students' abilities. For example: 20 – 8 =; 27 – 13 =; 67 – 49 =; 225 – 112 =; What is the difference between 58 and 75?; Erin had 45 drinks but she sold 27. How many did she have left? Have students record the equations and work out the answers. Discuss with students the different strategies they used to solve the equations.

WHOLE-CLASS INTRODUCTION

MISSING NUMBERS – SUBTRACTION

You will need: NTO 4.14 'Hundred Chart: FDP'

Write the equation 40 – □ = 20 on the board. Ask, 'What is the missing number? How could we work this out?' Discuss strategies with students and record them on the board. Repeat with another equation, e.g. 68 – □ = 21. Draw out the links between addition and subtraction and how they could be used to find the answer. Repeat with a number of examples.

INDEPENDENT TASKS

Note: Choose from Tasks 1, 2 or 3.

You will need: counters, empty containers (e.g. ice-cream containers), calculators, Student Book p. 113 'Missing Numbers – Subtraction'

TASK 1: FINDING THE DIFFERENCE

Give pairs of students counters and an empty container (e.g. an ice-cream container). Have students place a known number of counters under the upside-down container. Have students record the number. Then one of the pair removes a number of the counters. The other member of the pair counts the remaining counters and records an equation, e.g. 50 – □ = 25. Then this student solves the equation and checks the answer with their partner. Have students take turns. This activity could be varied by having students remove two lots of counters, with a viewing in between, and students recording two equations to solve.

TASK 2: INTERACTIVE TASK

Give students subtraction equations with missing numbers. Equations could include numbers in the hundreds or thousands, according to students' abilities, e.g. 1125 – □ = 498. Have students use calculators to help find the missing numbers. Give students a number of examples.

TASK 3: STUDENT BOOK p. 113 *'Missing Numbers – Subtraction'*

TEACHING GROUP

You will need: counters, poster paper

COUNTING BACK

- For students who require support, give them a random number of counters (more than 10). Have them count and record the number of counters. Then ask students to remove counters so they only have 10 left. Have them work out how many counters were taken away. Have students record this as an equation, e.g. 47 – □ = 10. Repeat a number of times, changing the number of counters remaining. Then redistribute the counters and have students repeat the activity.

VARYING EQUATIONS

- For students who require a challenge, give them an equation suited to their ability, e.g. 72 – ☐ = 35. Have students find the answer, then record all of the variations of the equation (fact family): 72 – 35 = 37, 72 – 37 = 35, 37 = 72 – 35, 35 = 72 – 37. This could also be extended to include addition: 35 + 37 = 72, 37 + 35 = 72, 72 = 35 + 37, 72 = 37 + 35. Have students organise their equations neatly on poster paper. Have students repeat this with a number of examples.

REFLECTION

Select from the following to suit your class and their learning outcomes:

- Review the answers to Student Book p. 113 'Missing Numbers – Subtraction'. Have students share strategies for finding missing numbers, particularly for Questions 3 and 5.
- Have students write a written reflection about the activity in Independent Tasks, Task 1. Have students describe how they felt and what they learned from the activity. Ask, 'What was the hardest equation? Which part did you like best?'
- Write a couple of examples on the board, e.g. 24 – ☐ = 17 and 62 = 98 – ☐. Have students discuss how they would find the missing numbers. Stress the importance of checking answers and draw out the links between addition and subtraction.

LESSON PLAN 3

TUNING IN

I AM THINKING OF A NUMBER

Pose this problem to students: 'I am thinking of a number. I add 10 to it, then I subtract 2. My answer is now 16. What number did I start from?' Read the question a number of times, then in groups have students try to find the starting number. (8)

WHOLE-CLASS INTRODUCTION

HOW DOES IT WORK?

Look at the problem from Tuning In. Work with students to discuss how they solved the question, and how it works (you can solve it by working backwards or working with the opposite operations). Give students another example: 'I am thinking of a number. I add 25, then subtract 5. My new number is 120. What was my starting number?' (100) Have students solve and discuss the strategies they used to find the answer.

INDEPENDENT TASKS

Note: Choose from Tasks 1, 2 or 3.

You will need: *PowerPoint*, Student Book p. 114 'I Am Thinking …'

TASK 1: I AM THINKING OF MY OWN NUMBER

Have pairs of students write 'I am thinking of a number' problems with as many steps as they like. The main criterion is that they must be able to find the answer. Have students type the problems up or present on a card, so a class set can be created.

TASK 2: INTERACTIVE TASK

Have students work independently on computers, using *PowerPoint* to create their own quiz. Their quiz must include one 'I am thinking of a number' problem, at least one missing number problem involving subtraction, and at least one missing number problem involving addition.

TASK 3: STUDENT BOOK p. 114 *'I Am Thinking …'*

TEACHING GROUP

You will need: modelling equipment (e.g. counters)

I AM THINKING OF SMALL NUMBERS

- For students who require support, repeat the 'I am thinking of a number' activity using smaller numbers, e.g. 'I am thinking of number, I add 3, and subtract 5. My new number is 2. What is my starting number?' Have students use modelling equipment to show what is happening: start with 2 counters then add 5 and subtract 3 to find the starting number. Have students check the answer. Repeat with a number of examples.

I AM THINKING OF MULTIPLICATION

- For students who require a challenge, have them write 'I am thinking of a number' problems including multiplication as well as addition and subtraction. Have them write at least three problems, then swap and check. This activity could be extended to include division.

REFLECTION

Select from the following to suit your class and their learning outcomes:

- Invite some students to share their 'I am thinking of a number' problems from Independent Tasks, Task 1, for the rest of the class to solve. Collect the problems not shared, and use as a maths warm-up each day.
- Have students share their *PowerPoint* quizzes from Independent Tasks, Task 2, with the rest of the class solving the problems. The quizzes not shared can be viewed at the beginning of lessons as revision.
- Have students share the problems that include multiplication from the Teaching Group 'I Am Thinking of Multiplication' for the rest of the class to try to solve.

Home Tasks

Select from the possible Home Tasks:

- Have students share their 'I am thinking of a number' problems from Lesson Plan 3, Independent Tasks, Task 1, with parents or carers. Have the adult solve the problem.
- Have students write two missing number equations – one using addition and one using subtraction – to bring to school and share.

Assessment

- Have students complete **Student Assessment p. 115**.
- Review with students **Assessment Task Card 4.28**.

During the three lessons:

- Collect the written reflections from Lesson Plan 2 as evidence of students' use of subtraction to find missing numbers.
- Collect copies of students' 'I am thinking of a number' problems from Lesson Plan 3, Independent Tasks, Task 1, to add to their portfolios.
- Collect students' *PowerPoint* quizzes from Lesson Plan 3, Independent Tasks, Tasks 2, as evidence of learning across the three areas of the unit.

Recommendations for Future Learning

Specific to Student Assessment p. 115; if the student is experiencing some difficulty:

Q 1–4 Provide similar problems based on smaller numbers, which enable the equation to be modelled with equipment. Have the student read the problems and identify the operations and the key elements. Have them attempt to write an equation, before attempting to solve.

Q 5 Revisit some of the strategies for finding missing numbers. The equations could be simplified to enable one step instead of two. Smaller numbers could also be used.

If the student has not achieved the recommended skills for this unit:

1. See **Assessment Task Card 4.28** for specific recommendations.
2. Review basic addition and subtraction processes.
3. Present problems and missing number sentences based on basic facts.
4. Have the student use modelling equipment to model the equations, and add and subtract elements accordingly.
5. Remind the student to check answers.
6. Review *Nelson Maths: Australian Curriculum NSW Year 3* Unit 21.

If the student has achieved the recommended skills and these skills are firmly established, consider:

1. Moving forward to *Nelson Maths: Australian Curriculum NSW Year 5* Unit 28.
2. Extending the student in any of the listed activities by working with larger numbers and increasing the use of multiplication and division operations.
3. Extending the student into problems with decimals.

Unit 29 Displaying Data

Statistics and Probability

Data MA2-18SP selects appropriate methods to collect data, and constructs, compares, interprets and evaluates data displays, including tables, picture graphs and column graphs

column graphs, data, many-to-one, picture graphs, tables

LESSON PLAN 1

TUNING IN

PICTURE GRAPHS

You will need: one-quarter of an A4 sheet of paper per student, coloured pencils/felt pens, large poster paper

Give each student one-quarter of an A4 sheet of paper. Have them draw their favourite animal. Use the pictures to create a class picture graph of students' favourite animals on a large piece of poster paper.

WHOLE-CLASS INTRODUCTION

KEY FEATURES AND INTERPRETING THE DATA

You will need: class graph from Tuning In

Using the class graph from Tuning In, add and point out the main features of the graph, e.g. arrows and labels on the axes, a heading. Ask questions to enable students to interpret the data, such as, 'What is the most popular animal? What is the least popular? What two animals had the same numbers?'

INDEPENDENT TASKS

Note: Choose from Tasks 1, 2 or 3.

You will need: poster paper, rulers, coloured pencils/felt pens, *Word* or *Kid Pix*, Student Book p. 116 'Picture Graphs'

TASK 1: ONE PICTURE = MANY VALUES

Provide students with some data, e.g. the number of vehicles in a car park: 40 cars, 10 vans/trucks, 7 motorbikes, 15 pushbikes, 3 scooters. Have students create a picture graph of the data on poster paper, where one picture represents five of that vehicle. So, for 40 cars, students will need eight pictures. Have students create a key on their graph to show this. Remind students to label the features on their graphs.

TASK 2: INTERACTIVE TASK

Ask, 'What is your favourite sport?' Quickly collate the data on the board. Then have students work independently on computers, using *Word* or *Kid Pix* to create picture graphs of the data.

TASK 3: STUDENT BOOK p. 116 *'Picture Graphs'*

TEACHING GROUP

You will need: coloured counters, LO: *L86 'Cassowary sanctuary'*

GRAPHING COUNTERS

- For students who require support, give them a handful of counters to sort into different coloured piles. Then have students line up the counters against a straight edge to create a picture graph. Ask students questions about the graph, e.g. 'Which is the most common coloured counter?' Have students record their graph on a sheet of paper. If necessary, students could trace around each of the counters.

LO: L86 'CASSOWARY SANCTUARY'

- For students who require a challenge, have them work in pairs on computers, using LO: *L86 'Cassowary sanctuary'*. Students work with mathematical concepts, such as shapes, area, fractions and graphs, to see how people can help to save cassowaries.

REFLECTION

Select from the following to suit your class and their learning outcomes:

- Have students present the graphs they made in Independent Tasks, Task 1. On the back of their graphs, have them write three statements about their data.
- Invite students to share their graphs from the Teaching Group 'Graphing Counters', explaining what they did to produce the graphs. Have the rest of the group comment about what results they can see, e.g. 'It seems like red counters were the most common in all of the graphs'.
- Review Student Book p. 116 'Picture Graphs' with students. Invite them to share their statements/observations about the graphs.

LESSON PLAN 2

TUNING IN

COLLECTING DATA

Divide the class into small groups of three or four. In the groups, have students discuss and brainstorm a question they would like to collect data about on the topic of the environment. Note: it may be appropriate to link the topic to a general studies theme or topic of interest at the school. Have students record the question.

WHOLE-CLASS INTRODUCTION

HOW TO COLLECT THE DATA?

You will need: question from Tuning In

Ask one group to share their question from Tuning In. Refine the question if needed. Ask, 'Who are we going to ask this question?' As a class, determine who the audience will be, e.g. the class, the year level, the Year 6 students. Then ask, 'How are we going to collect the data?' Collect students' ideas on the board. If appropriate, lead students towards a table or tally table. Share questions from other groups to ensure the questions are appropriate.

INDEPENDENT TASKS

Note: Choose from Tasks 1, 2 or 3.

You will need: poster paper, *Word* or *Excel*, Student Book p. 117 'Collecting Data'

TASK 1: COLLECTING THE DATA

Have students collect the data about their question. Students organise the data, perhaps as a table or list. Then on poster paper, groups show the data and a picture graph to illustrate the collected data.

TASK 2: INTERACTIVE TASK 

Have the groups work on computers, using *Word* or *Excel* to represent their data. Students may use *Excel* for the table and then *Word* for the picture graph. To extend the activity, have students represent their data as a column graph.

TASK 3: STUDENT BOOK p. 117 *'Collecting Data'*

TEACHING GROUP

You will need: sticky notes, hoops or chalk

VENN DIAGRAMS

- For students who require support, have them work as a group to create a Venn diagram on a topic of interest. Give each student a sticky note and have them write their name on it. Use hoops or chalk circles on the floor to make the Venn diagram, label each circle and then have each student stand in the respective circle. When satisfied students are in the right place, have them place the sticky note on the floor, and then leave the circles. Have students record the diagram on a sheet of paper, with the circles and then numbers for the total of sticky notes.

COLLECTING DATA FOR DIAGRAMS

- For students who require a challenge, have them work in pairs to create their own questions that could have data collected into a Venn diagram and 2-way table. Allow them to collect the data to see if their question was effective.

REFLECTION

Select from the following to suit your class and their learning outcomes:

- Have each group report on their question from Independent Tasks, Tasks 1 and 2, presenting their chart and digital representation. Ask, 'What did you learn about the topic by collecting your data?'
- Have students write an individual reflection about their learning of developing a question, collecting data, representing that data and then sharing it with the class.
- Review students' responses to Student Book p. 117 'Collecting Data'. Ensure students have represented the data correctly. Ask, 'What do you notice about the Venn diagram and the 2-way table?'

LESSON PLAN 3

TUNING IN

CREATING A COLUMN GRAPH

As a class, collect data about students' favourite fruit. Collate the data and represent it on the board as a picture graph. Then move to drawing boxes around the pictures to create the columns. Remove the pictures, leaving the columns. Discuss with students what the height of the columns represents.

WHOLE-CLASS INTRODUCTION

INTERPRETING THE COLUMN GRAPH

You will need: the graph from Tuning In

Discuss the main features of the graph from Tuning In, e.g. the arrows on the axes, the labels of the axes, the values of the heights of the columns. Spend time looking at how to determine the scale on the graph. Have students make some observations and comments about the graph, e.g. 'Apples are the most popular fruit'.

INDEPENDENT TASKS

Note: Choose from Tasks 1, 2 or 3.

You will need: BLM 17 '1 cm Grid Paper', LO: *L3512 'Bar chart'*, Student Book p. 118 'Column Graphs'

TASK 1: COLLECTING DATA FOR A COLUMN GRAPH

Have pairs of students determine a topic to collect some information about, e.g. favourite food, favourite football team. Have students devise their question and then collect the data and create a column graph of their data. BLM 17 '1 cm Grid Paper' could be used to provide structure for the graphs.

TASK 2: INTERACTIVE TASK

Have students work in their pairs from Task 1 on computers, using LO: *L3512 'Bar chart'*, which is a program that will enable them to plot a graph. Have students use the program to plot the data from their selected topic, comparing their final result to their hand-created graph.

TASK 3: STUDENT BOOK p. 118 *'Column Graphs'*

TEACHING GROUP

You will need: BLM 17 '1 cm Grid Paper', highlighter pens, rulers, coloured counters (20 per student)

COLUMN GRAPHS

- For students who require support, give them some simple data (according to abilities) to plot a column graph on BLM 17 '1 cm Grid Paper'. Work with number values between zero and 10 to allow for easy scaling. Make sure students use rulers to create clear lines and columns. Have students add features, e.g. arrows on axes, labels and headings. Then have them highlight the main features. Students can use this as a reference.

REPRESENTING COMPARISONS

- For students who require a challenge, have them guess what colours they might receive if they were given 20 counters. Students record the guesses in a table. Then give them a mix of 20 different-coloured counters. Have them record the actual colours and add this to the table. Give students BLM 17 '1 cm Grid Paper' and have them represent the two sets of data on the same axis.

REFLECTION

Select from the following to suit your class and their learning outcomes:

- Have students share their graphs from Independent Tasks, Tasks 1 and 2. Check that students have included the necessary features on the graph.
- Have pairs of students swap their graphs with different pairs, and then have the new pair write three observations from the graph.
- Have students reflect on the differences and similarities between their hand-drawn and computer-generated graphs. This could be completed as a table. Ask students which method they preferred and why.

Home Tasks

Select from the possible Home Tasks:

- Have students look for graphs in newspapers or magazines, cut out any examples and bring them to school. Have students identify the bar charts and picture graphs.
- Have students collect some data at home, e.g. the number of plates, cups and glasses. Have them represent their data in a column graph.

Assessment

- Have students complete **Student Assessment p. 119**.
- Review with students **Assessment Task Card 4.29**.

During the three lessons:

- Keep a copy of samples, such as students' graphs of vehicle data from Lesson Plan 1, Independent Tasks, Task 1, to add to their portfolio, as evidence of ability to create and interpret graphs.
- Collect copies of digital materials, such as graphs from Lesson Plan 2, to add to students' digital portfolios.
- Collect written reflections about data collection from Lesson Plan 2 as evidence of students' learning about displaying data, presenting their work and the ability to work in a group.

Recommendations for Future Learning

Specific to Student Assessment p. 119; if the student is experiencing some difficulty:

Q 1–2 Provide the student with a graph that has one-to-one correspondence of values, and have them answer questions about the graph/data.

Q 3 Have the student convert a graph of one-to-one correspondence, rather than many-to-one. The student could take the data from the picture graph, create a table of values and then construct the bar chart.

If the student has not achieved the recommended skills for this unit:

1. See **Assessment Task Card 4.29** for specific recommendations.
2. Review the basic construction of a graph, e.g. axes, labels, headings.
3. Have the student work with graphs of modelling equipment, then move to picture graphs and, finally, column graphs.
4. Have the student examine a range of graphs. Provide the student with a strategy of looking at axis labels to help determine what is being represented.
5. When the student is collecting data, provide structures such as tables or tally charts to help organise data.
6. Review *Nelson Maths: Australian Curriculum NSW Year 3* Unit 11.

If the student has achieved the recommended skills and these skills are firmly established, consider:

1. Moving forward to *Nelson Maths: Australian Curriculum NSW Year 5* Unit 29.
2. Extending the student in any of the listed activities by working with larger values and more data amounts.
3. Extending the student to consider where graphs could be used in real life.
4. Having the student develop the use of spreadsheet applications.

Unit 30 Interpreting Data

Statistics and Probability

Data MA2-18SP selects appropriate methods to collect data, and constructs, compares, interprets and evaluates data displays, including tables, picture graphs and column graphs

column graphs, data, many-to-one, picture graphs, tables

LESSON PLAN 1

TUNING IN

LOOKING AT DATA

You will need: data from the newspaper, e.g. the weather page

Show students a page of data from the newspaper, either online or a hard copy. The weather page could be used. Have students examine the page, and look at the information and how it is presented. Have them note five comments about the numbers, presentation, data and mathematics that they can see.

WHOLE-CLASS INTRODUCTION

USING THE DATA

You will need: comments and data from Tuning In

Have students share their comments from Tuning In. Ask questions about the data and what information it presents. Ask questions so students are using the data, e.g. 'What is the temperature expected to be on Thursday? What other ways can the data be organised?' Gather ideas and record these on the board.

INDEPENDENT TASKS

Note: Choose from Tasks 1, 2 or 3.

You will need: data source from Tuning In, poster paper, rulers, coloured pencils/felt pens, *Excel*, Student Book p. 120 'Graphing the Weather'

TASK 1: USING THE DATA

Have students re-express the data from Tuning In as a table or graph (or both). Have students decide on the headings and organisation of the table/graph. Remind students of elements of graphs and tables, e.g. straight lines for axes. Students present their work on poster paper.

TASK 2: INTERACTIVE TASK

Have students work independently on computers, using the data from Tuning In and Task 1 to create a graph of the information in *Excel*. Have students select which aspect they would like to graph and then allow them to explore the different graphing elements of *Excel*, e.g. different formats, different axes. Allow students to explore which of the different types of graphs best represent the data they have selected, e.g. a line graph for continuous data or a bar chart for discrete data.

TASK 3: STUDENT BOOK p. 120 *'Graphing the Weather'*

TEACHING GROUP

You will need: coloured counters or sticky notes, internet access

INTERPRETING DATA

- For students who require support, provide them with data for one week (or less, so they have fewer numbers to work with). Ask, 'What does this data show? Can we organise it another way?' Have students use counters or pictures on sticky notes to then model the graph. Have students verbalise what can be seen in the representations of the data.

VISITING THE WEATHER SITE

- For students who require a challenge, have them work on computers to visit the Bureau of Meteorology website, looking for the different ways that data is represented, e.g. tables, graphs, maps, charts, written descriptions. Have students make notes of what they find, and digitally collect the links or images of relevant examples and store in them a file so they can be shared.

REFLECTION

Select from the following to suit your class and their learning outcomes:

- Have students present their graphs and tables from Independent Tasks, Task 1. Ask students to explain why they selected the particular format. Ask, 'Does this type of graph best represent the data? Why?'
- Have students write a reflection on how they constructed their graph in Independent Tasks, Task 2, and why they selected the particular format. Ask, 'Does this type of graph best represent the data? Why?
- Invite students to share the different data representations that they found on the Bureau of Meteorology website in the Teaching Group 'Visiting the Weather Site'. Have students show examples using the IWB. Select one of the examples and have students interpret the information.

LESSON PLAN 2

TUNING IN

IN THE SPORTS SECTION

You will need: the sports section of a newspaper

Have students examine the data they can find in the sports section of a newspaper. On the board, create a list of the data, graphs and tables that are found.

WHOLE-CLASS INTRODUCTION

1 REPRESENTS MORE

Draw a symbol on the board, e.g. ● = 5.

Ask, 'What could this mean?' (It could mean: 1 soccer ball = 5 balls; or 1 soccer ball = 5 goals; or 1 soccer ball = 5 people who like soccer as a sport.) Draw another example and discuss why one symbol often means more than one item.

INDEPENDENT TASKS

Note: Choose from Tasks 1, 2 or 3.

You will need: newspapers, scissors, glue, poster paper, *Word*, Student Book p. 121 'Interpreting Data'

TASK 1: NEWSPAPER HUNT

Have students look through the newspaper for data representations – this could be from any section. Have them cut out an example and stick it on poster paper. Students write a comment about whether this is an effective representation of the data, and why. Have students then write five statements about their data on the poster paper.

TASK 2: INTERACTIVE TASK

Have students reinterpret their data and devise symbols where one item represents more than one data value, e.g. ● = 5 points. Once they determine their symbols, have students represent their data in a table on *Word*. This activity could be extended by having students represent their data as a picture graph.

TASK 3: STUDENT BOOK p. 121 *'Interpreting Data'*

TEACHING GROUP

You will need: sticky notes, Unifix blocks, Student Book p. 121 'Interpreting Data', a digital camera, LO: *L2625 'Media report: junk food'*

UNIFIX GRAPH

- For students who require support, give them a copy of the data from Student Book p. 121 'Interpreting Data'. As a group, have them find the totals and read the information in the table. Then allocate one person per vegetable and have them find that number of Unifix blocks, e.g. 50 green blocks for lettuce. Then have students create the graph by placing their Unfix blocks in a column and lining them up beside each other to make the graph. Have students add labels to the graph with sticky notes, then take a digital photo of the graph. To extend the activity, students could translate the graph to the Student Book page.

LO: L2625 'MEDIA REPORT: JUNK FOOD'

- For students who require a challenge, have them work on computers, using LO: *L2625 'Media report: junk food'* to help an editor check an article drafted for a magazine that contains a column graph showing recent consumption of junk food by different age groups.

REFLECTION

Select from the following to suit your class and their learning outcomes:

- Have students share the selected data they found in the newspaper in Independent Tasks, Task 1. Ask, 'Why did you select this data? What did you learn from looking at the data?'
- Have students share their tables from Independent Tasks, Task 2, with symbols representing the data. Ask, 'What does your data show? Tell us one interesting fact you discovered.'
- Have students from the Teaching Group 'Unifix Graph' share the Unifix graph of the Student Book page. Discuss with students why 50 items are used when there is a column but only five when there is a picture.

LESSON PLAN 3

TUNING IN

CREATING A GRAPH

You will need: BLM 17 '1 cm Grid Paper' or poster paper, rulers

Provide students with a table of data that has only headings for each column, but not an overall heading or theme. Have pairs of students draw either a picture graph or a column graph. Students could draw up their graph on BLM 17 '1 cm Grid Paper' or on poster paper.

WHOLE-CLASS INTRODUCTION

SHARING THE GRAPHS

You will need: the graph from Tuning In

Under their graphs from Tuning In, have students write five comments about what they learned. Ask them also to identify what the information could represent. Have students share their graphs, comments and ideas. Discuss as a group.

INDEPENDENT TASKS

Note: Choose from Tasks 1, 2 or 3.

You will need: BLM 52 'Insect Column Graph', poster paper, scissors, glue, rulers, *Excel*, *Word*, Student Book p. 122 'At the Beach'

TASK 1: COLLECTING DATA FROM A COLUMN GRAPH

Give students a copy of BLM 52 'Insect Column Graph' and have them stick it on poster paper. Then have students create a table of values from the graph. Students develop their own heading for the graph, and then write three facts from the data.

TASK 2: INTERACTIVE TASK

Have students work in pairs to create a newspaper article that contains data. Topics could be allocated, or they could select their own, referring to newspapers to develop ideas. Have students use their preferred method to produce their article, e.g. students may want to use *Excel* to produce the graph, and then *Word* to create the article.

TASK 3: STUDENT BOOK p. 122 ***'At the Beach'***

TEACHING GROUP

You will need: Unifix blocks, BLM 17 '1 cm Grid Paper', poster paper, *Excel*, *Word*

SHOWING THE DATA

- For students who require support, give them some Unifix blocks to use in creating a column graph. Have students then record the graph on BLM 17 '1 cm Grid Paper'. When complete, have students break apart their Unifix and create a table of values/data. Ask students to look at the similarities/differences between the data and the graph.

MAKING ERRORS

- For students who require a challenge, have them create a graph and table that don't quite match, to create a 'Find the differences' puzzle. These could be completed on poster paper or electronically.

REFLECTION

Select from the following to suit your class and their learning outcomes:

- Have students share their findings from the data collected in Independent Tasks, Task 1. Ask, 'Why did you think the graph was about …? How did you know there were 30 bees? What other ways could we have represented the information?'

- Print students' newspaper articles from Independent Tasks, Task 2, and create a class newspaper. Share the different articles with students, asking them to explain how they came up with the different ideas.
- Copy the puzzles created by students in the Teaching Group 'Making Errors'. Have the other students in the class find the differences.

Home Tasks

Select from the possible Home Tasks:

- Select a graph from a newspaper or other source, delete the heading and give a copy to each student. At home, have students create a matching table of data, write three facts and state a heading for the data. Have them share the activity with parents or carers.
- Have students take home their newspaper article from Lesson Plan 3, Independent Tasks, Task 2, to read to parents or carers. Have a parent/carer write a comment about something they learned or thought was interesting about the article.

Assessment

- Have students complete **Student Assessment p. 123**.
- Review with students **Assessment Task Card 4.30**.

During the three lessons:

- Collect a copy of students' work in reinterpreting data into a graph from Lesson Plan 1, Independent Tasks, Task 2, as evidence of their understanding of interpreting data.
- Collect copies of digital materials, such as the tables created in Lesson Plan 2, Independent Tasks, Task 2, as an example of the understanding of data representation. Add these materials to students' digital portfolios.
- Collect students' charts from Lesson Plan 3, Independent Tasks, Task 1, as evidence of students' ability to interpret data and add this to their portfolios.

Recommendations for Future Learning

Specific to Student Assessment p. 123; if the student is experiencing some difficulty:

Q 1 Give the student graphs from different sources, e.g. the newspaper, mask the headings and have the student suggest headings.

Q 2 Give the student simple graphs and ask questions about them. When the student is confident, have them suggest questions for others to answer, based on the data provided.

Q 3 Review the links between tables of values and graphs. Have the student practise drawing/creating one from the other where 1 symbol = 1 data value.

If the student has not achieved the recommended skills for this unit:

1. See **Assessment Task Card 4.30** for specific recommendations.
2. Review the basic construction of a graph, e.g. axes, labels.
3. Review the structure of a table, e.g. rows, columns, labels.
4. Have the student practise answering questions and writing questions about data sets. If this is too challenging, the student could match questions to provided data sets.
5. Have the student look at media sources, e.g. newspapers and the internet, for different representations of data.
6. Review *Nelson Maths: Australian Curriculum NSW Year 3* Unit 11.

If the student has achieved the recommended skills and these skills are firmly established, consider:

1. Moving forward to *Nelson Maths: Australian Curriculum NSW Year 5* Unit 30.
2. Extending the student in any of the listed activities by working with larger values and more data amounts.
3. Having the student work with complex graphs from different media sources.
4. Having the student critique different data representations.

Unit 31 Money

Number and Algebra

Addition and subtraction (money) MA2-5NA uses mental and written strategies for addition and subtraction involving two-, three-, four- and five-digit numbers

Fractions and decimals (money) MA2-7NA represents, models and compares commonly used fractions and decimals

cents, change, currency, decimal systems, dollars, money, place value, purchases, rounding

LESSON PLAN 1

TUNING IN

WHAT COULD IT BE?

You will need: a selection of coins

Pose this problem to students: 'If I had 70 cents in my pocket, what coins could it be?' Have students brainstorm possibilities. Gather ideas and record these on the board. Repeat again with: 'If I had $5 in my pocket, what coins could it be?'

WHOLE-CLASS INTRODUCTION

CHANGE

You will need: a variety of notes and coins, poster paper

Pose some money questions, e.g. 'My items cost $1.50. How much change would I receive from $2? $5?' Have students share how to find the answer. Model with the coins and notes. Explore on the board how to calculate. Repeat a number of times. Then pose a question that has an amount that requires rounding, e.g. $1.97. Create and display a chart reminding students about rounding, e.g. round up if it ends with 7, 8, 9.

INDEPENDENT TASKS

Note: Choose from Tasks 1, 2 or 3.

You will need: BLM 54 'Coins', scissors, dice, calculators, envelopes/plastic pockets, internet access, Student Book p. 124 'Change'

TASK 1: FIND THE CHANGE

Give each pair of students a copy of BLM 54 'Coins' to cut up. Have students turn over the coins and mix them up. One student rolls a dice, collects that number of coins, adds up the value and records this on a sheet of paper. Then the student finds the change from the nearest dollar, and records this as well. Students take turns. They could stick the coins on the sheet of paper each time, until all of the coins are used. Note: students who require assistance could use a calculator. At the completion of the activity, leftover coins could be stored in an envelope or plastic pocket.

TASK 2: INTERACTIVE TASK

Have students work independently on computers to explore the website of the Royal Australian Mint (where coins are made). If students look under the 'Design and Products' section, they will find information about all of the decimal coins made, how many are produced each year and what one- and two-cent coins looked like. Have students record five interesting facts from the website.

TASK 3: STUDENT BOOK p. 124 *'Change'*

TEACHING GROUP

You will need: BLM 54 'Coins', BLM 55 'Banknotes', LO: *L4159 'Diffy'*, calculators

PRACTISING WITH COINS AND NOTES

- For students who require support, provide them with copies of BLM 54 'Coins' and BLM 55 'Banknotes'. Have students practise adding up money, using the coins and notes on the BLMs, and recording the totals (use real money if there is enough). Remind students of skills such as grouping to 10. When students are confident, set amounts to make such as $1.75. Again, have students record these.

LO: L4159 'DIFFY'

- For students who require a challenge, have them work on computers, using the money option on LO: *L4159 'Diffy'* to complete the puzzle. Students may a calculator for some of the calculations.

REFLECTION

Select from the following to suit your class and their learning outcomes:

- Have students share the five interesting facts they found on the Royal Australian Mint website in Independent Tasks, Task 2. Ask, 'When did they stop making one- and two-cent coins? Why do you think that happened?'
- Review the answers to Student Book p. 124 'Change'. Have students explain how they found the answers to particular questions.
- Have the class brainstorm different ways of making $5.60. Collect their ideas on the board.

LESSON PLAN 2

TUNING IN

A HANDFUL OF COINS

You will need: BLM 54 'Coins' or real coins

Cut up multiple copies of BLM 54 'Coins' or use real coins. Have students grab a handful of coins, then nominate amounts to make with the coins. Then have them find the change from the nearest dollar, then the nearest $10, listing the notes and coins that would make up that change.

WHOLE-CLASS INTRODUCTION

WHAT SHALL I BUY?

You will need: catalogues, calculators, NTO 4.3 'Calculator'

Photocopy, enlarge and cut out some items from catalogues (including prices) that may be of interest to students, e.g. toys, sporting equipment, electronic equipment. Attach these to the board. Have students select three of the items they would like, write down the name of each item and the price and then find the total of the items. Allow students to use a calculator if they need help with the calculation. Remind students how to enter the decimals on the calculator (NTO 4.3 'Calculator' could be used). For other students, have them complete the calculation, and then check their answers with calculators.

INDEPENDENT TASKS

Note: Choose from Tasks 1, 2 or 3.

You will need: catalogues or access to the internet, scissors, glue, poster paper, calculators, copies of the school canteen menu, *Excel*, Student Book p. 125 'Off to the Movies'

TASK 1: LET'S GO SHOPPING

Give students access to catalogues or the internet. Tell students they have $100 to spend. They are to buy one thing for school, one thing to eat and then they can spend the rest however they like. They are to buy at least five different items. Students cut out (or print out) the item, stick it on poster paper, find the total of the items and the change from the $100. Students can use calculators to assist in their calculations.

TASK 2: INTERACTIVE TASK

Give students a copy of the school canteen menu. Have them select three different combinations of lunches – each combination must include a drink and a snack, as well as a main item, e.g. a sandwich. In *Excel*, have students create three different lunch totals, where the item and the amount are listed in separate columns and the sum function is used to find the total.

TASK 3: STUDENT BOOK p. 125 *'Off to the Movies'*

TEACHING GROUP

You will need: copies of the school canteen menu, paper bags, BLM 54 'Coins', BLM 55 'Banknotes', calculators, furniture catalogues, internet access

TOTALLING MY LUNCH

- For students who require support, give them a copy of the school canteen menu. Have them circle five items they would like to eat, then write their order on a paper bag. Spend time talking about layout – students could draw one large and one small column for the descriptions and the amounts. Have them estimate the total cost of the lunch, and place the relevant money in the bag using BLM 54 'Coins' and BLM 55 'Banknotes'. Then have them check the totals using calculators and review their estimations.

COST OF A BEDROOM

- For students who require a challenge, have them work in pairs to find the cost of furnishing a new bedroom. Give students catalogues or access to the internet (the websites of large furniture stores such as Ikea would be useful). Have them create a list of items and find the total. They may wish to include images.

REFLECTION

Select from the following to suit your class and their learning outcomes:

- Have students share their shopping choices from Independent Tasks, Task 1. Ask, 'What was your item for school? Why did you select that food item? How much change did you have left?'
- Have students write a written reflection about their shopping in Independent Tasks, Task 1. Have them explore: how they selected their items, when they knew they had spent too much and how they calculated the total of their shopping and the change.
- Have students share their cost of going to the movies and lunch from Student Book p. 125 'Off to the Movies'. How much change did mothers receive, or was there money owed?

LESSON PLAN 3

TUNING IN

IN THE PAST

You will need: an IWB or data projector, *R11234 'Decimal Currency Advertisement, 1965'* (this can be found on the econtent repository http://econtent.thelearningfederation.edu.au/ec/p/home)

Ask students what Australia's currency was before decimal currency was introduced. On the IWB or data projector, show them the footage from *R11234 'Decimal Currency Advertisement, 1965'*. Discuss.

WHOLE-CLASS INTRODUCTION

OTHER COUNTRIES

You will need: NTO 4.18 'Money from Other Countries'

Brainstorm currencies from other countries. Draw on students' knowledge; ideas can also be stimulated from NTO 4.18 'Money from Other Countries'. Talk about what is different and what is the same, e.g. use of decimal systems.

INDEPENDENT TASKS

Note: Choose from Tasks 1, 2 or 3.

You will need: internet access, poster paper, LOTE activity on money from the econtent repository http://econtent.thelearningfederation.edu.au/ec/p/home, Student Book p. 126 'In China'

TASK 1: MONEY POSTERS

Allocate pairs or small groups of students a country's currency to investigate. Have them produce a poster showing information about the money, and some simple calculations.

TASK 2: INTERACTIVE TASK

Have students work independently on computers, using a selection from the econtent repository http://econtent.thelearningfederation.edu.au/ec/p/home. For example, *'Super! 1, 2, 3'* has an option to listen to descriptions of items and prices and match them to a price tag in a range of languages.

TASK 3: STUDENT BOOK p. 126 *'In China'*

TEACHING GROUP

You will need: NTO 4.18 'Money from Other Countries', real money from other countries (if possible), newspapers or access to the internet

ONE COUNTRY'S MONEY

- For students who require support, give them a country's currency to explore further – this could be the LOTE country. Spend some time looking at the currency, either on NTO 4.18 'Money from Other Countries' or on the internet. Have students complete some simple calculations with the money, e.g. adding the value of the coins together and adding the value of the notes. If possible, use the real money and have students look at the differences and similarities to Australian currency.

THE VALUE OF MONEY

- For students who require a challenge, have them find the value of other countries' money in Australia, e.g. the US dollar, the euro, the Japanese yen. Have students collect the values, perhaps in a table. Information could be sourced from a newspaper, the TV news or the internet.

REFLECTION

Select from the following to suit your class and their learning outcomes:

- Have students share their money posters from Independent Tasks, Task 1, with the group. Have them explain their calculations. Students from the Teaching Group 'One Country's Money' could also share.
- Have students share what they discovered from completing the LOTE activity in Independent Tasks, Task 2.
- Invite students from the Teaching Group 'The Value of Money' to share what they discovered about the value of different currencies in Australia. Ask, 'Do you think you could buy more in this country than Australia? Why?'

Home Tasks

Select from the possible Home Tasks:

- Have students share their money posters from Lesson Plan 3, Independent Tasks, Task 1, with parents or carers and explain what they learned about the other country's money system.
- Invite students to bring in any foreign money they have to share with the class.
- Have students create a birthday wish list with three items on it. With the help of parents and carers, students find out the individual and total cost of the items. Have them bring the list to school and share.

Assessment

- Have students complete **Student Assessment p. 127**.
- Review with students **Assessment Task Card 4.31**.

During the three lessons:

- Keep a copy of samples of students' calculations with money, such as their shopping items from Lesson Plan 2, Independent Tasks, Task 1, for their portfolios.
- Collect copies of digital materials, such as students' *Excel* lunch totals from Lesson Plan 2, Independent Tasks, Task 2, for their digital portfolios.
- Make note of students who completed the extension tasks and of interesting discoveries they made.

Recommendations for Future Learning

Specific to Student Assessment p. 127; if the student is experiencing some difficulty:

Q 1 Have the student practise finding totals of sets of coins and notes using BLM 54 'Coins' and BLM 55 'Banknotes'. Start with simple combinations, building to more coins and notes.

Q 2–3 Review the processes of finding change. This would include finding the total first, and then either subtracting from the final amount and finding the difference or counting up to that final amount.

Q 3 Have the student practise finding totals using the calculator. Make sure the student is entering the data correctly with the decimal point, and then reading and recording the amounts correctly.

Q 4 Revisit some of the elements of Lesson Plan 3, particularly the student-generated posters and NTO 4.18 'Money from Other Countries'.

If the student has not achieved the recommended skills for this unit:

1. See **Assessment Task Card 4.31** for specific recommendations.
2. Review basic addition strategies to aid with calculations, e.g. grouping to 10 and counting by 5s and 10s.
3. Practise rounding with the student.
4. Review the two strategies of counting up to the total to find the change or using subtraction to find the difference.
5. Make sure the student is using the calculator correctly, entering the decimal values and recording values correctly.
6. Review *Nelson Maths: Australian Curriculum NSW Year 3* Unit 32.

If the student has achieved the recommended skills and these skills are firmly established, consider:

1. Moving forward to *Nelson Maths: Australian Curriculum NSW Year 5* Unit 31.
2. Extending the student by using values expressed with zeros, e.g. $0.55 or $1.05.
3. Having the student complete addition calculations manually, even with multiple numbers, then check answers on a calculator.

Unit 32 Word Problems

Number and Algebra

Patterns and algebra MA2-8NA generalises properties of odd and even numbers, generates number patterns, and completes simple number sentences by calculating missing values

division, multiplication, number sentence, word problem

LESSON PLAN 1

TUNING IN

AT THE BEGINNING

Present a word problem, such as: 'Albert collected 12 apples from one tree, 14 apples from another and 13 apples from the third tree. How many apples had Albert collected?' Have students solve the problem on a sheet of paper. Then present students with an equation, e.g. 23 – 18 =. Have students solve the equation, and then write a word problem for the number sentence.

WHOLE-CLASS INTRODUCTION

EXPLORING WORD PROBLEMS

You will need: solutions from Tuning In

Discuss with students how they solved the problem in Tuning In, look at different strategies and gather ideas and record these on the board. Emphasise strategies, such as underlining key terms and identifying the appropriate operation. If necessary, provide another example. Then have students share their answers and written word problems for the number sentence 23 – 18 =. Discuss which examples were interesting and what elements made good word problems. These could be collected and made into a class chart.

INDEPENDENT TASKS

Note: Choose from Tasks 1, 2 or 3.

You will need: *Word*, Student Book p. 128 'Word Problems'

TASK 1: WRITING WORD PROBLEMS

Give pairs of students a set of addition and subtraction equations, according to their abilities. Have students write each equation, solve it and then, on another sheet of paper, write a matching word problem. When finished, have students swap their word problems and try to solve them. This activity could be extended by having students type up their word problems.

TASK 2: INTERACTIVE TASK

Have students work independently on computers, using *Word* to create a 2 × 2 table to use as a think board. Give students an addition or subtraction equation (according to their abilities). Students put the equation in the first cell, then in the next cell, they write a matching word problem. In the third cell, they place images to show the equation in a 'hands-on' or modelled format. In the final cell, students write a hint for someone trying to solve their equation, e.g. 'Use a number line starting at 15 and counting by 2s' or 'Use a place-value chart to organise the equation'. This activity could be extended by having students complete the inverse of their equation, e.g. if they had 15 + 16 =, they could repeat the activity for 31 – 15 =.

TASK 3: STUDENT BOOK p. 128 *'Word Problems'*

TEACHING GROUP

PRACTICE QUESTIONS

- For students who require support, give them simple word problems based on addition and subtraction, e.g. 'Ester has 5 apples and 3 oranges. How much fruit does she have altogether?' Review strategies for finding the number sentence, e.g. underline the numbers, identify the operation and write the number sentence. If necessary, provide materials to help solve the equation. Give students a number of problems to solve, according to their ability. Continue discussing with students how they developed the number sentence.

DECIMAL WORD PROBLEMS

- For students who require a challenge, give them decimal word problems to solve, such as: 'Maxine measured 2 lengths of wood – one was 1.25 m and the other was 1.73 m. What was the total length of the wood?' and 'The weights of 3 potatoes were 0.75 g, 0.78 g and 0.95 g. What was the total weight of the potatoes?' This activity could be altered by providing the decimal equations and having students write matching word problems.

REFLECTION

Select from the following to suit your class and their learning outcomes:

- Have students share their word problems from Independent Tasks, Task 1. Alternatively, these could be collated to form a worksheet for students to complete.
- Have students share their think boards from Independent Tasks, Task 2, either electronically or on hard copy. Invite them to share their images and hints and have the rest of the class solve the problems.
- Reflect with students on how to solve word problems. Create a chart of hints to display in the classroom.

LESSON PLAN 

TUNING IN

IN THE CAR PARK

You will need: poster paper

Pose the following problem: 'There are 5 vehicles in the car park. How many wheels might there be?' Have pairs or small groups of students brainstorm different solutions to the problem. Have them collect their ideas on poster paper and share their responses with the group.

WHOLE-CLASS INTRODUCTION

IT MEANS MULTIPLICATION

You will need: poster paper

Discuss what operations students used to solve the problem in Tuning In. Draw out that they could have used addition or multiplication. Ask, 'How did you know what operation to use? Why?' On poster paper, brainstorm all the terms that mean addition, and then all the terms that mean multiplication.

INDEPENDENT TASKS

Note: Choose from Tasks 1, 2 or 3.

You will need: craft sticks or match sticks, LO: *L8156 'What's the problem: sports: level 1'*, Student Book p. 129 'Multiplication Problems'

TASK 1: STICKY PROBLEMS

Give pairs of students craft sticks or match sticks. Have them create a square with the sticks. Then have them create another square, then another one. Have students write the pattern and then develop the multiplication rule, i.e. number of squares × 4 = number of sticks. Using their rule, have students work out how many sticks are in 20 squares and how many squares could be made from 200 sticks. Repeat the activity, investigating triangles. This activity could be extended by having students investigate different shapes, or examining what type of pattern is formed when the multiple shapes are joined.

TASK 2: INTERACTIVE TASK

Have students work independently on computers, using LO: *L8156 'What's the problem: sports: level 1'*. Students select worded mathematical problems about sports and identify the correct answer for each step in the problem-solving process with the help of a panel.

TASK 3: STUDENT BOOK p. 129 ***'Multiplication Problems'***

TEACHING GROUP

You will need: modelling equipment (e.g. counters, Unifix blocks), LO: *L8154 'What's the problem: nature: level 1'*

ADDING VISUALS

- For students who require support, provide a number of multiplication equations, e.g. 10 × 2 =. Give students modelling equipment, e.g. counters or Unifix blocks, to model the equation and find the answer. Then have them write a word problem, e.g. 'I have 10 groups of 2 counters. How many counters do I have?' Repeat a number of times. This task could be extended by having students look at both representations of an equation, e.g. 10 × 2 = and 2 × 10 =. Students could explore how this looks with the equipment and the subtle difference with the language in the written equation, i.e. 2 groups of 10 counters.

LO: L8154 'WHAT'S THE PROBLEM: NATURE: LEVEL 1'

- For students who require a challenge, have them work independently on computers, using LO: *L8154 'What's the problem: nature: level 1'*. Students select worded mathematical problems about nature, and identify the correct answer for each step in the problem-solving process. Note: there are also Levels 2 and 3 versions of this activity to extend students further.

REFLECTION

Select from the following to suit your class and their learning outcomes:

- Have students share their results from their stick investigations in Independent Tasks, Task 1. Ask, 'What patterns did you find? What was different between each of the shapes?'
- Have students reflect on the Learning Object activities. Ask, 'How did the panel members help you solve the problems? What was difficult about the activity? Did the process get easier after you had tried it a number of times?'
- Show students an array (NTO 4.10 'Arrays' could be used). Have them write a word problem that would match the array. Have students share their problems.

LESSON PLAN 3

TUNING IN

DIVISION PROBLEMS

Provide students with the two different types of division problems (partition and quotition). For example: 'Twenty books were put on 5 shelves, with an equal number on each shelf. How many books were on each shelf?' (partition) 'Mary wanted to make some 4 m ribbons from a length of 21 m. How many ribbons could she make?' (quotition) Have students solve the problems. Encourage students to discuss the remainder in the second problem.

WHOLE-CLASS INTRODUCTION

DIFFERENT TYPES

You will need: answers from Tuning In

Discuss with students the similarities and differences between the problems in Tuning In. Say, 'Explain how you solved the problem.' You could draw diagrams on the board and have students identify the division equation. Draw out the links between multiplication and division. Ask students to explain how they knew to use division or repeated subtraction.

INDEPENDENT TASKS

Note: Choose from Tasks 1, 2 or 3.

You will need: modelling equipment (e.g. counters or Unifix blocks), BLM 17 '1 cm Grid Paper', LO: *L8155 'What's the problem: planets: level 1'*, Student Book p. 130 'Division Word Problems'

TASK 1: MODELLING DIVISION PROBLEMS

Allocate pairs of students some division equations to complete based on their tables facts, according to students' abilities. Have students solve the equations, using modelling equipment. This may be counters or Unifix blocks, or arrays using BLM 17 '1 cm Grid Paper'. Have students record the equation, the modelling and the answer. Repeat with a number of equations. To extend the activity, have students also find the related multiplication equations.

TASK 2: INTERACTIVE TASK

Have students work independently on computers, using LO: *L8155 'What's the problem: planets: level 1'*. Students solve word problems about planets, and identify the correct answer for each step in the problem-solving process.

TASK 3: STUDENT BOOK p. 130 ***'Division Word Problems'***

TEACHING GROUP

You will need: poster paper

TERMS THAT MEAN

- For students who require support, have them divide a piece of poster paper into four sections. Label each section: +, −, × and ÷. In each section, have students record all the terms they know that mean addition, subtraction, multiplication and division. Have them also record an example of both a number sentence and a word problem for each section.

DIVISION WITH REMAINDERS

- For students who require a challenge, give them an open-ended task, e.g. 'A number is divided by 6 and leaves a remainder of 3. What might this number be?' Have pairs of students explore the problem and see if they can come to a rule.

REFLECTION

Select from the following to suit your class and their learning outcomes:

- Have students share their modelled division problem from Independent Tasks, Task 1, explaining their model and how it represents the equation.
- Invite some students to share the word problems they wrote on Student Book p. 130 'Division Word Problems'. Look at the different types of questions produced for the same equation.
- Have students from the Teaching Group 'Division with Remainders' share their work. Ask, 'What strategies did you use to solve the problem? Was there a rule you could come up with?'

Home Tasks

Select from the possible Home Tasks:

- Have students divide a sheet of paper into four. Have them write an addition equation in one section, then a subtraction, multiplication and division equation in each of the other sections. Have students take the paper home and have parents or carers or a friend come up with word problems for their equations. Students may like to ask different family members to give them an equation for each section. Have students bring this to school to share with the class.
- Have students write three word problems based on their home, e.g. 'I have 9 chickens, 1 rabbit, 2 dogs and 3 ducks. How many pets do I have altogether?' Have students bring these to school to share.

Assessment

- Have students complete **Student Assessment p. 131**.
- Review with students **Assessment Task Card 4.32**.

During the three lessons:

- Collect students' work from Lesson Plan 3, Independent Tasks, Task 1, as an example of their understanding of division, modelling of the equation and creating word problems.
- Collect copies of digital materials, such as the think boards from Lesson Plan 1, Independent Tasks, Task 2, for students' digital portfolios.
- Collect students' work from the Lesson Plan 3 Teaching Group 'Terms That Mean', as evidence of students' understanding of the four operations. Add this to the student portfolios.

Recommendations for Future Learning

Specific to Student Assessment p. 131; if the student is experiencing some difficulty:

Q 1 Have the student model the equations to solve. Then have the student record the modelling, leading the verbalising and then the recording of a related number sentence.

Q 2 Select one element of the think board for the student to complete. Alternatively, provide the student with the equation or the operation to use, so the task is not so open-ended.

If the student has not achieved the recommended skills for this unit:

1. See **Assessment Task Card 4.32** for specific recommendations.
2. Use modelling equipment or strategies such as arrays and number lines to solve equations.
3. Emphasise the link between addition and subtraction, and multiplication and division to aid in the process of solving equations.
4. Have the student practise skills, such as underlining/highlighting terms, identifying operations and using diagrams, to aid in the solving of equations.
5. Use modelling equipment to help create situations for developing word problems.
6. Review *Nelson Maths: Australian Curriculum NSW Year 3* Unit 15.

If the student has achieved the recommended skills and these skills are firmly established, consider:

1. Moving forward to *Nelson Maths: Australian Curriculum NSW Year 5* Unit 6.
2. Extending the student by using equations that include decimals or units of measurement.
3. Having the student use equations that have more than one step and/or more than one operation.

Unit 33 Patterns

Measurement and Geometry

Two-dimensional space MA2-15MG manipulates, identifies and sketches two-dimensional shapes, including special quadrilaterals, and describes their features

patterns, shapes, symmetry

LESSON PLAN 1

TUNING IN

FINDING SYMMETRY

You will need: playdough, an assortment of shape cutters, plastic knives

Give students some playdough (this can be made from flour, cream of tartar, salt and water – there are many recipes available on the internet). Have them use the cutters to cut out a range of shapes to investigate which are symmetrical. Students can check by cutting the shape in half and placing it on top of the other half. The activity could be extended by having students look for different lines of symmetry or recording their findings.

WHOLE-CLASS INTRODUCTION

SYMMETRY

Have students brainstorm the meaning of symmetry. Collect the different ideas on the board. Ask, 'How do we know if something is symmetrical? How do we know a shape is symmetrical?' Collect a list of shapes that students consider symmetrical. Consider a triangle and the different forms, e.g. equilateral as opposed to a scalene triangle.

INDEPENDENT TASKS

Note: Choose from Tasks 1, 2 or 3.

You will need: poster paper, *Word*, Student Book p. 132 'Symmetrical Numbers and Shapes'

TASK 1: SYMMETRICAL LETTERS

Ask students to investigate which letters of the alphabet (both upper- and lower-case) are symmetrical and which are not. Have them present the information as a table on poster paper. This activity could be extended by having students come up with words where all the letters are symmetrical.

TASK 2: INTERACTIVE TASK

Have students work independently on computers, using *Word* to examine letters in different fonts to see which are symmetrical and which are not. Does the font affect whether or not the letters are symmetrical? It may be easier if students enlarge letters and use lines to test their ideas. Have them present their findings.

TASK 3: STUDENT BOOK p. 132 *'Symmetrical Numbers and Shapes'*

TEACHING GROUP

You will need: LO: *L7803 'Finding symmetry: one line: city'*, BLM 62 'Symmetrical Pictures'

LO: L7803 'FINDING SYMMETRY: ONE LINE: CITY'

- For students who require support, have them work independently on computers, using LO: *L7803 'Finding symmetry: one line: city'* to explore a futuristic city and find shapes with one line of symmetry. Note: there are various versions of this Learning Object to cater for different abilities.

SYMMETRICAL PICTURES

- For students who require a challenge, have them complete BLM 62 'Symmetrical Pictures'. This activity could be further extended with an online version at www.haelmedia.com/OnlineActivities_txh/mc_txh4_001.html.

REFLECTION

Select from the following to suit your class and their learning outcomes:

- Have students share their charts from Independent Tasks, Task 1. Check to see if all students have the same answers. Discuss any that were trickier than others. Display the work.
- Have students share their work from Independent Tasks, Task 2. Ask, 'Did enlarging the letters make it easier to consider if the letters were symmetrical? Did changing the font affect your results?'
- Invite students to share their work from the Teaching Group 'Symmetrical Pictures', either on hard copy or electronically.

LESSON PLAN 2

TUNING IN

SYMMETRY IN ARTWORK

You will need: examples of artworks that exhibit symmetry

Look at some examples of artworks that include symmetry. There are many sites, including: http://www.fun-stuff-to-do.com/geometric-shapes.html as well as examples on the econtent repository http://econtent.thelearningfederation.edu.au/ec/p/home. Other examples could include motifs in Central Asian textiles, Tibetan artefacts, Indian lotus designs and symmetry in Yolngu or Central and Western Desert art (as stated in the Australian Curriculum). If sufficient examples are found, an art display could be held.

WHOLE-CLASS INTRODUCTION

SYMMETRY IN DESIGN

You will need: artwork examples from Tuning In

Brainstorm with students what they saw in the artwork, how symmetry played a part in the design and where examples might be found, e.g. clothing, book covers, wall art, tiling.

INDEPENDENT TASKS

Note: Choose from Tasks 1, 2 or 3.

You will need: artwork examples from Tuning In, BLM 17 '1 cm Grid Paper', BLM 56 '5 mm Grid Paper', BLM 57 'Tessellation Patterns 1', BLM 58 'Tessellation Patterns 2', coloured pencils/felt pens, Student Book p. 133 'Describing Art'

TASK 1: DESIGNING A PATTERN

Allow students to select from BLM 17 '1 cm Grid Paper', BLM 56 '5 mm Grid Paper', BLM 57 'Tessellation Patterns 1' and BLM 58 'Tessellation Patterns 2'. Have them use their selected BLM to create a design or pattern based on symmetry. Allow students to use the artwork examples from Tuning In as inspiration.

TASK 2: INTERACTIVE TASK

Allow students to select their preferred computer program to create a design or pattern based on symmetry. Students may wish to complete more than one design.

TASK 3: STUDENT BOOK p. 133 *'Describing Art'*

TEACHING GROUP

You will need: pattern blocks, a digital camera, BLM 17 '1 cm Grid Paper', BLM 56 '5 mm Grid Paper', BLM 57 'Tessellation Patterns 1', BLM 58 'Tessellation Patterns 2', coloured pencils/felt pens

CREATING WITH PATTERN BLOCKS

- For students who require support, give them an assortment of pattern blocks and have them create patterns. Look at the patterns that contain symmetry and talk about the different shapes used. Take digital photos of students' work. This activity could be extended by having students complete the pattern creation as a group.

DESIGNING A RUG

- For students who require a challenge, give them a limited choice of shapes, e.g. squares and triangles, and have them create a design for a rug of a certain size. Allow students to use any tools they choose, e.g. BLMs or computers, to complete their creation.

REFLECTION

Select from the following to suit your class and their learning outcomes:

- On the back of their patterns from Independent Tasks, Task 1, have students write a description of what the pattern is and how symmetry is evident. Display students' work.
- Collect and display students' designs from Independent Tasks, Task 2, perhaps as a *PowerPoint* display. Have students talk about their own design. Display the work on the IWB or on a digital photo frame.
- Have students share their creations from the Teaching Group 'Creating with Pattern Blocks', explaining where the symmetry is and naming the shapes they have used.
- Have students share their designs from the Teaching Group 'Designing a Rug', explaining how they solved the design challenge.

LESSON PLAN 3

TUNING IN

FOLLOWING DIRECTIONS

You will need: digital cameras

In pairs or groups of three, have students use digital cameras to find patterns with symmetry around the school. This may include patterns in brick works, tiles, artwork and in art displays, both inside and outside the classroom. Have students take turns to use the camera.

WHOLE-CLASS INTRODUCTION

SELECTING PHOTOS

You will need: the images from Tuning in

Have students work in their pairs/groups to upload their photos, and select five images to show the rest of the group. Students need to be able identify the shapes in the patterns, as well as the symmetry and also why they selected the image. Have students share their selections and their explanations.

INDEPENDENT TASKS

Note: Choose from Tasks 1, 2 or 3.

You will need: coloured paper, scissors, LO: *L3540 'Pentominoes'*, Student Book p. 134 'Not Patterns'

TASK 1: WITH COLOUR

Give pairs of students a selection of different-coloured paper, and have them create their own pattern, similar to tiles or a mosaic. Discuss with students economical and efficient methods for cutting out the paper. Have students plan their design before cutting.

TASK 2: INTERACTIVE TASK

Have students work individually on computers, using LO: *L3540 'Pentominoes'* to explore the creation of pentominoes and patterning.

TASK 3: STUDENT BOOK p. 134 ***'Not Patterns'***

Note: there is an opportunity to discuss that sometimes the opposite of something is used in maths to show (or prove) something else. In this case, students are encouraged *not* to create patterns, illustrating their understanding.

TEACHING GROUP

You will need: LO: *L7788 'Tessellate decorate: three shapes'*

LO: L7788 'TESSELLATE DECORATE: THREE SHAPES'

- For students who require support, have them work individually on computers, using LO: *L7788 'Tessellate decorate: three shapes'* to decorate a room in a house with patterns made of rhombuses, squares and equilateral triangles.

IN THE LIBRARY

- For students who require a challenge, have them visit the library to look for books and materials that have examples of patterns and symmetry in art. This may include picture books. Have students consider materials that come from different cultures.

REFLECTION

Select from the following to suit your class and their learning outcomes:

- Have students write a commentary about their digital photos from Tuning In, the selection process and how they worked as a pair or group. Have them consider what they might do differently next time, what they learned about their own learning, and their partner or other group members.
- Have students share Student Book p. 134 'Not Patterns' and explain how they solved the problem.
- Have students from the Teaching Group 'In the Library' share what they found from their library search, explaining why they identified particular materials.

Home Tasks

Select from the possible Home Tasks:

- Give students a copy of BLM 17 '1 cm Grid Paper', BLM 56 '5 mm Grid Paper', BLM 57 'Tessellation Patterns 1' and BLM 58 'Tessellation Patterns 2' and have them create a symmetrical pattern. Have students bring their patterns to school and display their work.
- Have students look for different symmetrical patterns at home, either inside or outside. Students may be able to take a digital photo, or they could sketch or describe the pattern.

Assessment

- Have students complete **Student Assessment p. 135**.
- Review with students **Assessment Task Card 4.33**.

During the three lessons:

- Collect digital presentations of symmetrical letters and fonts from Lesson Plan 1, Independent Tasks, Task 1, as examples of students' understanding of symmetry and their organisation in presenting content. Add these to students' digital portfolios.
- Collect the patterns designed by students as evidence of their understanding of symmetry.
- Collect students' written commentary from Lesson Plan 3 about their digital photos and working in pairs/groups.

Recommendations for Future Learning

Specific to Student Assessment p. 135; if the student is experiencing some difficulty:

Q 1 Review the concept of symmetry. Review the lines of symmetry in letters and numbers, and basic shapes. The student could complete paper cutting exercises to see if a shape has symmetry.

Q 2–3 Revisit the idea of creating patterns with different materials. Examine examples of artwork and other objects, and have students articulate the shapes and patterns that they see, and if the pattern has symmetry.

If the student has not achieved the recommended skills for this unit:

1. See **Assessment Task Card 4.33** for specific recommendations.
2. Revisit basic shapes with the student, matching names to shapes to aid identification and articulation. Materials such as NTO 4.9 '2D Shapes and 3D Objects' could be used.
3. Work with hands-on materials, e.g. pattern blocks, paper shapes, playdough shapes, so the student can manipulate shapes to create patterns.
4. Activities such as the video game Tetris and shape-based puzzles could also help the student.
5. Review *Nelson Maths: Australian Curriculum NSW Year 3* Unit 29.

If the student has achieved the recommended skills and these skills are firmly established, consider:

1. Having the student develop more intricate patterns with greater detail.
2. Moving forward to *Nelson Maths: Australian Curriculum NSW Year 5* Unit 32.
3. Providing the student with conditions related to area and perimeter to create links to the real world. For example, for a specified 1 m^2 area, design a pattern using a particular set of tiles.

Year 4: Assessment Task Card

Odd and Even Numbers
Resources: A4 paper

1 Give each student a sheet of paper. Have them write an addition equation that has an even answer.
2 Have students write a subtraction equation that has an odd answer.
3 Have students state if the answer to 10 × 5 = is odd or even.
4 Using their knowledge of odd and even numbers, have students state whether the answer to 451 x 43 = is odd or even.
5 For students who require extension, have them write a division equation that has an even answer.

Whole numbers
MA2-4NA applies place value to order, read and represent numbers of up to five digits
Addition and subtraction
MA2-5NA uses mental and written strategies for addition and subtraction involving two-, three-, four- and five-digit numbers
Multiplication and division
MA2-6NA uses mental and informal written strategies for multiplication and division

Year 4: Assessment Task Card

4.1

Odd and Even Numbers TARGETED ASSESSMENT

If the student is experiencing difficulty:

Q1 Review the addition of odd and even numbers with the 'Dividing Counters' activity from Lesson Plan 1, Teaching Group.

Q2 Review the subtraction of odd and even numbers with the 'Addition and Subtraction with Counters' activity from Lesson Plan 2, Teaching Group.

Q3–4 Use NTO 4.3 'Calculator' and NTO 4.4 'Number Line: Patterns' and look at patterns with the multiplication equations and answers.

If the student has demonstrated an understanding beyond the skills, consider:

Q5 Extending the difficulty of the division equations and consolidating the understanding. More time could be spent on LO: *L589 'Musical Number Patterns: Music Maker'*.

Whole numbers
MA2-4NA applies place value to order, read and represent numbers of up to five digits
Addition and subtraction
MA2-5NA uses mental and written strategies for addition and subtraction involving two-, three-, four- and five-digit numbers
Multiplication and division
MA2-6NA uses mental and informal written strategies for multiplication and division

Year 4: Assessment Task Card

4.2

Unit 2

Numbers to Tens of Thousands

Resources: for each student – A4 paper and A3 paper, scissors, glue

1 Give each student a piece of A4 paper. Have them divide and cut it into quarters (four pieces).
2 On one quarter, have students write a 5-digit number between 30 000 and 40 000.
3 On another quarter, have students write 45 263 in words.
4 On another quarter, have students write a 5-digit number that begins with 6.
5 On the last quarter, have students write a number less than 20 000.
6 For students who require extension, have them order the quarters from smallest to largest, then paste them in order on an A3 sheet,

Whole numbers
MA2-4NA applies place value to order, read and represent numbers of up to five digits

Year 4: Assessment Task Card

4.2

Unit 2

Numbers to Tens of Thousands

TARGETED ASSESSMENT

If the student is experiencing difficulty:

Q2–4 Review what a 5-digit number is. Revisit the place-value chart.
Q3 Revisit the 'Missing Words' activity from Lesson Plan 1, Teaching Group.
Q5–6 Use NTO 4.4 'Number Line: Patterns' to review location of numbers, comparing larger and smaller. It may also be useful to revisit the place-value chart.

If the student has demonstrated an understanding beyond the skills, consider:

Q6 Extending the complexity of the ordering of the numbers, by having the student create a scaled number line.

Whole numbers
MA2-4NA applies place value to order, read and represent numbers of up to five digits

Year 4: Assessment Task Card

Place Value
Resources: A4 paper

1 Give each student a sheet of paper. Have them write three 5-digit numbers.
2 Have students select one of their numbers and expand it.
3 Have students select another one of their numbers, then expand and rename the number.
4 With the third number, have students identify the value of the thousands. Ask students to write how they worked this out.
5 For students who require extension, have them select one of their numbers and draw a representation of it.

Whole numbers
MA2-4NA applies place value to order, read and represent numbers of up to five digits

Year 4: Assessment Task Card

Place Value TARGETED ASSESSMENT

If the student is experiencing difficulty:

Q1 Review what a 5-digit number is. Revisit the place-value chart.

Q2–3 Have the student use a number expander or NTO 4.7 'Number Expander' to expand numbers and rename them.

Q4 Review the place-value chart (NTO 4.5 'Place-Value Mat: Millions' could be used). Review the different columns and what they mean.

If the student has demonstrated an understanding beyond the skills, consider:

Q5 Challenging the student to find all representations of a particular number, i.e. digit, words, expanded form, renamed and modelled with different equipment, e.g. MAB.

Whole numbers
MA2-4NA applies place value to order, read and represent numbers of up to five digits

Year 4: Assessment Task Card

Length and Temperature

Resources: A4 paper, one ruler per student, five Unifix blocks per student, NTO 4.8 'Thermometer'

1. Give each student a sheet of paper and a ruler. Ask them to draw and label a 10 cm line on the paper.
2. Ask them to build a tower of five Unifix blocks, measure the length of the tower and record the length on the paper.
3. Ask students to list three things that would be measured in centimetres.
4. Ask students to list three things that would be measured in metres.
5. Using NTO 4.8 'Thermometer', set a temperature and have students record this on their page.
6. For students who require extension, set two temperatures on NTO 4.8 'Thermometer' and have students record the two temperatures and find the difference.

Length

MA2-9MG measures, records, compares and estimates lengths, distances and perimeters in metres, centimetres and millimetres, and measures, compares and records temperatures

Year 4: Assessment Task Card

Length and Temperature TARGETED ASSESSMENT

If the student is experiencing difficulty:

Q1–2 Review how to use a ruler with the student, revisiting the number scale and the units on the ruler.

Q3–4 Use the *PowerPoint* presentations students created in Lesson Plan 2, Independent Tasks, Task 2, and discuss which types of objects are measured in centimetres and which are measured in metres. The student could also revisit measuring objects in the classroom.

Q5 Revisit the use of the thermometer and the reading of the scale, using NTO 4.8 'Thermometer'.

If the student has demonstrated an understanding beyond the skills, consider:

Q6 Having the student work with more complicated differences, perhaps moving into negative temperatures.

Length

MA2-9MG measures, records, compares and estimates lengths, distances and perimeters in metres, centimetres and millimetres, and measures, compares and records temperatures

Year 4: Assessment Task Card

Mass and Capacity

Resources: A4 paper, three objects to weigh (e.g. a shoe, a ball and a pencil case), three containers, measuring equipment (e.g. scales)

1 Give each student a sheet of paper. Ask them to write down the order of the mass of three objects from lightest to heaviest.

2 Ask students to estimate the mass of each object.

3 Have students measure and record the mass of each object.

4 Ask students to list three things that would be measured in litres.

5 Ask students to list three things that would be measured in millilitres.

6 For students who require extension, show them three containers and ask them to write an explanation of how to find the capacity of the three containers using only water.

Volume and capacity

MA2-11MG measures, records, compares and estimates volumes and capacities using litres, millilitres and cubic centimetres

Mass

MA2-12MG measures, records, compares and estimates the masses of objects using kilograms and grams

Year 4: Assessment Task Card

4.5

Mass and Capacity TARGETED ASSESSMENT

If the student is experiencing difficulty:

Q1–3 Review how to estimate the mass of an object, then weigh it on scales. Check that the student is reading the scales correctly, and waiting for the scales to settle at zero before weighing.

Q4–5 Review the difference between millilitres and litres. Using containers marked with the different units, have the student measure 2 mL and 2 L, and compare.

If the student has demonstrated an understanding beyond the skills, consider:

Q6 Having the student work with conversions between each of the unit sets, i.e. millilitres and litres, grams and kilograms.

Volume and capacity

MA2-11MG measures, records, compares and estimates volumes and capacities using litres, millilitres and cubic centimetres

Mass

MA2-12MG measures, records, compares and estimates the masses of objects using kilograms and grams

Year 4: Assessment Task Card

4.6

Unit 6

Number Sequences: 3s, 6s and 9s

Resources: A4 paper

1. Give each student a sheet of paper. Have them write the 3s number sequence starting at zero, for 10 numbers.
2. Then have students write the 3s number sequence starting at 10, for 10 numbers.
3. Give students the numbers 12, 18, 24, … and have them complete the next five numbers of the sequence.
4. Ask students to write how they worked out the numbers in the sequence in Question 3.
5. Ask students to write a 9s number sequence from any starting point, for six numbers.
6. For students who require extention, ask them to write an explanation of the links between the 3s, 6s and 9s number sequences.

Multiplication and division

MA2-6NA uses mental and informal written strategies for multiplication and division

Year 4: Assessment Task Card

4.6

Unit 6

Number Sequences: 3s, 6s and 9s TARGETED ASSESSMENT

If the student is experiencing difficulty:

Q1–2 Review the 3s number sequence with the aid of Hundred Charts and number lines, either hard copy or electronic (see NTO 4.1 'Hundred Chart' and NTO 4.4 'Number Line: Patterns').

Q3–4 Review the 6s number sequence with the aid of Hundred Charts and number lines, either hard copy or electronic (see NTO 4.1 'Hundred Chart' and NTO 4.4 'Number Line: Patterns'). This could be further scaffolded with the concept of AFL scoring to engage interest.

Q5 Review the 9s number sequence, with the list and observations made. This could be further supported with the use of Hundred Charts and number lines, either hard copy or electronic (see NTO 4.1 'Hundred Chart' and NTO 4.4 'Number Line: Patterns').

If the student has demonstrated an understanding beyond the skills, consider:

Q6 Having the student extend the examination of links to include the 12s number sequence.

Multiplication and division

MA2-6NA uses mental and informal written strategies for multiplication and division

Year 4: Assessment Task Card

Number Sequences: 4s, 8s and 7s
Resources: A4 paper

1 Give each student a sheet of paper and a number sequence according to their ability. Have them identify the sequence.

2 Give students the numbers 24, 32, 40, … and have them complete the next four numbers in the sequence.

3 Have students write an explanation of how they worked out the next four numbers.

4 Have students complete a 7s number sequence, giving five numbers in the sequence from any starting point.

5 Ask students to write a 4s number sequence starting at 100 and continuing for the next eight numbers.

6 For students who require extension, have them start at 100 and provide the 7s number sequence backwards.

Multiplication and division
MA2-6NA uses mental and informal written strategies for multiplication and division

Year 4: Assessment Task Card 4.7

Number Sequences: 4s, 8s and 7s TARGETED ASSESSMENT

If the student is experiencing difficulty:

Q1 Review the different number sequences with the aid of Hundred Charts and number lines, either hard copy or electronic (see NTO 4.1 'Hundred Chart' or NTO 4.4 'Number Line: Patterns').

Q2–3 Review the 8s number sequence with the aid of Hundred Charts and number lines, either hard copy or electronic (see NTO 4.1 'Hundred Chart' and NTO 4.4 'Number Line: Patterns'). This could be further supported by looking at the 4s, and using the concept of doubling.

Q4 Review the 7s number sequence, with students' *PowerPoint* presentations from Lesson Plan 3, Independent Tasks, Task 2. This could be further supported with the use of Hundred Charts and number lines, either hard copy or electronic (see NTO 4.1 'Hundred Chart' and NTO 4.4 'Number Line: Patterns').

Q5 Review the 4s number sequence with the aid of Hundred Charts and number lines (extending beyond 100), either hard copy or electronic (see NTO 4.1 'Hundred Chart' and NTO 4.4 'Number Line: Patterns').

If the student has demonstrated an understanding beyond the skills, consider:

Q6 Having the student work with a range of number sequences from any starting point, either forwards or backwards.

Multiplication and division
MA2-6NA uses mental and informal written strategies for multiplication and division

Year 4: Assessment Task Card

4.8

Unit 8

Regular Shapes

Resources: BLM 17 '1 cm Grid Paper', BLM 18 'Regular and Irregular Shapes', flat blocks

1 Give each student a copy of BLM 17 '1 cm Grid Paper' and either a regular or irregular shape (these could be from BLM 18 'Regular and Irregular Shapes'). Have students trace the shape on the 1 cm grid paper.

2 Have students record answers to:

a Is your shape regular or irregular?

b Why is your shape regular or irregular?

c What is the area of your shape?

d Draw an irregular shape that is larger in area than your shape.

3 For students who require extension, have them draw a regular shape to a specified size, e.g. 20 cm^2.

Two-dimensional space

MA2-15MG manipulates, identifies and sketches two-dimensional shapes, including special quadrilaterals, and describes their features

Year 4: Assessment Task Card

4.8

Unit 8

Regular Shapes TARGETED ASSESSMENT

If the student is experiencing difficulty:

Q2a–b Review the difference between regular and irregular shapes. Discuss strategies for identifying if side lengths and angles are equal.

Q2c–d Have the student trace a shape on grid paper and find the area of the shape. Observe how the student counts the squares. Talk about the importance of units. Have the student find a shape that is larger in area than the first one, then trace it and compare the two.

If the student has demonstrated an understanding beyond the skills, consider:

Q3 Challenging the student to draw shapes of a particular area. Have the student practise drawing five-, six- and eight-sided shapes accurately with the aid of 1 cm grid paper.

Two-dimensional space

MA2-15MG manipulates, identifies and sketches two-dimensional shapes, including special quadrilaterals, and describes their features

Year 4: Assessment Task Card 4.9

Unit 9

2D Shapes and 3D Objects

Resources: BLM 17 '1 cm Grid Paper', 2D shapes (flat blocks), 3D objects, BLM 32 'Isometric Dot Paper'

1. Give each student a copy of BLM 17 '1 cm Grid Paper' and a number of 2D shapes. Have them trace around the different shapes and identify the name of each shape, the number of edges and the number of vertices.
2. Have students create a compound shape with the 2D shapes and record this, including a written description of which common 2D shapes make up the compound shape.
3. Ask students to choose a 3D object and to sketch it freehand or on isometric dot paper.
4. For students who require extension, have them create and record a compound shape according to specified criteria, e.g. having six sides and using four different blocks.

Three-dimensional space

MA2-14MG makes, compares, sketches and names three-dimensional objects, including prisms, pyramids, cylinders, cones and spheres, and describes their features

Two-dimensional space

MA2-15MG manipulates, identifies and sketches two-dimensional shapes, including special quadrilaterals, and describes their features.

Year 4: Assessment Task Card 4.9

Unit 9

2D Shapes and 3D Objects — TARGETED ASSESSMENT

If the student is experiencing difficulty:

Q1 Review the names of the different shapes. NTO 4.9 '2D Shapes and 3D Objects' could be used as a matching activity. Review the terms 'edges' and 'vertices' with the student.

Q2 Review the term 'compound shape' with the student. Have the student practise making compound shapes with two blocks, then three, then four and so on. Review effective recording techniques with the student.

Q3 Review drawing a 3D object by copying it from a 2D drawing (as on Student Book p. 38 'Drawing 3D Objects').

If the student has demonstrated an understanding beyond the skills, consider:

Q4 Challenging the student to 'pull apart' simple compound shapes, leading to more complex shapes such as tangram pictures.

Three-dimensional space

MA2-14MG makes, compares, sketches and names three-dimensional objects, including prisms, pyramids, cylinders, cones and spheres, and describes their features

Two-dimensional space

MA2-15MG manipulates, identifies and sketches two-dimensional shapes, including special quadrilaterals, and describes their features.

Year 4: Assessment Task Card

Multiplication Facts (Times Tables)
Resources: A4 paper, two 10-sided dice for each student

1 Give each student a sheet of paper and two 10-sided dice. Have students roll the dice and record the numbers, and repeat this five times. Then have students multiply the pairs of numbers together.

2 Ask students to produce six multiplication equations that equal 24.

3 Ask students to draw four arrays to represent the total of 12.

4 Have students show the multiplication equation 7×2 on a number line.

5 For students who require extension, have them find the total cost of three balls at $9 each and explain how they did this.

Multiplication and division
MA2-6NA uses mental and informal written strategies for multiplication and division

Year 4: Assessment Task Card

4.10

Multiplication Facts (Times Tables) TARGETED ASSESSMENT

If the student is experiencing difficulty:

Q1 Review multiplication facts using BLM 3 'Tables Chart 1' and BLM 4 'Tables Chart 2'.

Q2 Review links between tables sets. This could be supported with the posters created by students in Lesson Plan 3, Independent Tasks, Task 1, or by looking at the links on BLM 3 'Tables Chart 1' and BLM 4 'Tables Chart 2'.

Q3 Review the concept and representation of arrays. Use the activities from Lesson Plan 1.

Q4 Use NTO 4.4 'Number Line: Patterns' to review the representation of tables on a number line.

If the student has demonstrated an understanding beyond the skills, consider:

Q5 Providing more complex tables or applying the student's knowledge to word problems.

Multiplication and division
MA2-6NA uses mental and informal written strategies for multiplication and division

Year 4: Assessment Task Card

Multiplication Facts and Related Division Facts
Resources: A4 paper

1 Give each student a sheet of paper and have them draw an array for 4 × 3.
2 Ask students to list as many multiplication equations as they can related to the array.
3 Ask students to list as many division equations as they can related to the array.
4 Provide students with the equation 6 × 5 = ? Ask students to solve it and then to give a related division equation.
5 For students who require extension, have them find how many movie tickets were bought, if the total cost was $72 and each ticket cost $9. Have students explain how they did this.

Multiplication and division
MA2-6NA uses mental and informal written strategies for multiplication and division

Year 4: Assessment Task Card

4.11

Multiplication Facts and Related Division Facts TARGETED ASSESSMENT

If the student is experiencing difficulty:

Q1 Review what an array is, what it looks like and how to draw it. The student could work with counters to model this.

Q2 Use a physical array to discuss multiplication equations. Look at rows and columns. Also look at equations that are not as obvious such as x1, leading to the fact families.

Q3 Review the concept that the inverse of multiplication is division. Then look at the equations from Question 2 and invert them. Alternatively, the student could return to tables charts such as BLM 3 'Tables Chart 1' and BLM 4 'Tables Chart 2'.

Q4 This could be reviewed using tables charts (e.g. BLM 3 'Tables Chart 1' and BLM 4 'Tables Chart 2') or the multiplication grids produced by students in Lesson Plan 1, Independent Tasks, Task 1.

If the student has demonstrated an understanding beyond the skills, consider:

Q5 Providing more word-based problems, or equations working with tables up to × 12, or tables based on multiples of 10 or 100.

Multiplication and division
MA2-6NA uses mental and informal written strategies for multiplication and division

Year 4: Assessment Task Card

Unit 12

Mapping

Resources: BLM 17 '1 cm Grid Paper'

1 Give each student a copy of BLM 17 '1 cm Grid Paper'. Ask them to draw an island shape on the grid, then add the following: a lake, a hut, a beach, a treehouse and a volcano.

2 Have students create a legend to add the following: five trees; roads between the lake, hut, treehouse and volcano; three quicksand pits.

3 Have students create a scale for their map.

4 Ask students to write a description of how they would travel between the lake, the treehouse and the volcano.

5 Have students add an 'X' for the treasure on their map, and write a description of how to get from the beach to the treasure.

6 For students who require extension, have them find the distance (using their scale) from the beach to the treasure.

Position

MA2-17MG uses simple maps and grids to represent position and follow routes, including using compass directions

Year 4: Assessment Task Card

Unit 12

Mapping TARGETED ASSESSMENT

If the student is experiencing difficulty:

Q2 Review what a legend is. Examine a number of maps with different legends. Revisit looking at the different features of a map and finding the common identifiers.

Q3 Review the concept of scale. Revisit maps to see the effect of scale. Make comparisons of normal-sized objects and relate on a smaller scale.

Q4–5 Have the student practise working with simple directions, e.g. taking five steps forwards/backwards, and then relating to objects around them. Have the student complete the movements to get a 'feel' for directions.

If the student has demonstrated an understanding beyond the skills, consider:

Q6 Challenging the student with mapping pathways and then finding the distance using the scale on the map.

Position

MA2-17MG uses simple maps and grids to represent position and follow routes, including using compass directions

Year 4: Assessment Task Card

Addition and Subtraction

Resources: A4 paper, four dice for each student

1 Give each student a sheet of paper and four dice. Have students roll the dice and record four 4-digit numbers. Note: this activity can be modified by using fewer dice.

2 Have students order the four numbers from smallest to largest. Then have students find the difference between the largest and the smallest numbers.

3 Ask students to add the two remaining numbers.

4 Have students then roll two dice and decide to either add or find the difference between the two numbers. Then have students make a pattern by multiplying the answer by 10 each time.

5 For students who require extension, have them add or subtract two of their initial numbers on a number line.

Addition and subtraction

MA2-5NA uses mental and written strategies for addition and subtraction involving two-, three-, four- and five-digit numbers

Year 4: Assessment Task Card

4.13

Addition and Subtraction

TARGETED ASSESSMENT

If the student is experiencing difficulty:

Q1 Review what a 4-digit number is.

Q2 Review what the term 'difference' means. Remind the student to write the largest number first. Review some of the strategies used for subtraction in Lesson Plan 2.

Q3 Review the concept of addition. Review some of the strategies from Lesson Plan 1. Concrete materials could also be used, e.g. MAB.

Q4 Review the patterns of increasing by a factor of 10 each time. Begin with simple equations and patterns and build to ones that are more complex.

If the student has demonstrated an understanding beyond the skills, consider:

Q5 Providing more word-based problems, or equations using numbers with more than four or five digits.

Addition and subtraction

MA2-5NA uses mental and written strategies for addition and subtraction involving two-, three-, four- and five-digit numbers

Year 4: Assessment Task Card

Perimeter and Area
Resources: BLM 17 '1 cm Grid Paper', ruler, flat shapes (e.g. pattern blocks), rulers

1 Give each student a copy of BLM 17 '1 cm Grid Paper' and a flat shape. Have students find the perimeter of the shape and record it on the BLM using their preferred method.
2 Have students find the area of the shape and record it.
3 Provide students with a second shape and have them make a compound shape and trace onto their grid.
4 Have them find the perimeter and area of the second shape.
5 Have students identify the shape of the largest area.
6 For students who require extension, provide them with an area, and have them construct a compound shape with this area.

Length
MA2-9MG measures, records, compares and estimates lengths, distances and perimeters in metres, centimetres and millimetres, and measures, compares and records temperatures
Area
MA2-10MG measures, records, compares and estimates areas using square centimetres and square metres

Year 4: Assessment Task Card

Perimeter and Area TARGETED ASSESSMENT

If the student is experiencing difficulty:

Q1–2 Review finding perimeter with simple shapes and then irregular shapes. Revisit the Teaching Group activities from Lesson Plan 1.

Q3–4 Review the process of counting squares to find the area of a shape. Review the possibility of dividing the shape into familiar shapes, e.g. squares and rectangles, recording the areas and then adding.

Q5 Areas can be compared by cutting out and overlaying one shape on another. Remind the student to examine units as well.

If the student has demonstrated an understanding beyond the skills, consider:

Q6 Challenging the student with more complex shapes, e.g. triangles.

Length
MA2-9MG measures, records, compares and estimates lengths, distances and perimeters in metres, centimetres and millimetres, and measures, compares and records temperatures
Area
MA2-10MG measures, records, compares and estimates areas using square centimetres and square metres

Year 4: Assessment Task Card

Multiplication and Division Strategies
Resources: A4 paper

1 Give each student a sheet of paper and three multiplication equations, e.g. 4 × 6 =, 5 × 7 =, 10 × 9 =. Have students write the reversed equivalent equations, i.e. 6 × 4 =, 7 × 5 =, 9 × 10 =.

2 Have students draw the arrays for one pair of the equations (e.g. 4 × 6 and 6 × 4).

3 Give students a number, e.g. 4, and have them double it and double it again, then write the related multiplication equations.

4 Give students another number, e.g. 20, and have them halve it, and write the related division equations.

5 For students who require extension, have them draw an array of 4 × 5 and then have them 'hide' half of it.

Multiplication and division
MA2-6NA uses mental and informal written strategies for multiplication and division

Year 4: Assessment Task Card

Multiplication and Division Strategies TARGETED ASSESSMENT

If the student is experiencing difficulty:

Q1 Review the concept of commutativity, where the order the equation is written does not matter. Use materials such as BLM 3 'Tables Chart 1' and BLM 4 'Tables Chart 2' to find equations that are reversed and equal.

Q2 Practise drawing arrays (NTO 4.10 'Arrays' can be used). Model different arrays and make the links to the multiplication equations.

Q3–4 Review the concepts of doubling and halving. Concrete materials can be used to model the process, with the student physically doubling or halving sets of materials.

If the student has demonstrated an understanding beyond the skills, consider:

Q5 Having the student use the ideas of arrays, doubling and halving to create their own equations and questions or problems.

Multiplication and division
MA2-6NA uses mental and informal written strategies for multiplication and division

Year 4: Assessment Task Card

More Multiplication and Division Strategies

Resources: A4 paper, two 10-sided dice per student, counters

1 Give each student a sheet of paper and two 10-sided dice. Have students roll the two dice and record the two numbers, then write two multiplication and two division equations using those numbers.

2 Have students record all of the factors of the rolled multiplied dice.

3 Have students multiply the multiplication equation by 10 and 100.

4 Show the student 11 counters. Say, 'Show me how you can use these to solve 11 ÷ 3.'

Multiplication and division

MA2-6NA uses mental and informal written strategies for multiplication and division

Year 4: Assessment Task Card

4.16

More Multiplication and Division Strategies TARGETED ASSESSMENT

If the student is experiencing difficulty:

Q1 Review multiplication using BLM 3 'Tables Chart 1' and BLM 4 'Tables Chart 2' or BLM 30 'Multiplication Grid'. Review the process of division with the multiplication grid to find answers. Have the student practise writing equations.

Q2 Review factors by looking at all of the multiplication equations with the same answers. This could be completed using BLM 3 'Tables Chart 1' and BLM 4 'Tables Chart 2' or BLM 30 'Multiplication Grid'.

Q3 Review the concepts of multiplying by 10 on the place-value chart and with calculators. Repeat with multiplying by 100.

Q4 Show five counters and ask, 'How many groups of three can we make? How many are left over?'

Multiplication and division

MA2-6NA uses mental and informal written strategies for multiplication and division

Year 4: Assessment Task Card

Volume and Capacity

Resources: various containers, measuring jugs, interlocking centimetre cubes

1 Show three different containers and ask the student to order the capacities from smallest to largest. You may also choose to have the student justify the responses by means of a practical activity.

2 Show a 250 mL marked container and a 1 L marked container. Ask, 'How many of the small container would fill the large one?'

3 Ask the student how they arrived at the answer to Question 2.

4 Ask the student to make a box-shaped model that has one layer and a volume of 12 cm^3.

5 Ask the student to repeat the activity in Question 4 with a model of a different length.

6 Ask, 'If you put a second layer on the first model, what would the volume be?' Repeat for the second model, if necessary.

Volume and capacity

MA2-11MG measures, records, compares and estimates volumes and capacities using litres, millilitres and cubic centimetres

Year 4: Assessment Task Card

Volume and Capacity TARGETED ASSESSMENT

If the student is experiencing difficulty:

Q1 Use containers that have their capacities marked.

Q2–3 Reinforce the method for finding out by means of a practical activity.

Q4 Ask the student to join the cubes in a line and describe the volume.

Q5 Repeat the activity in Question 4 and then ask the student to change the model into two lines of six. Establish that there are no more and no fewer cubes and ask, 'What is the volume?'

Q6 Change the models in Questions 4 and 5 so that they are each 6 cm by 2 cm. Ask, 'What would the volume be if you put one on top of the other?' Allow the student to count the cubes if necessary.

Volume and capacity

MA2-11MG measures, records, compares and estimates volumes and capacities using litres, millilitres and cubic centimetres

Year 4: Assessment Task Card

Decimals to Two Decimal Places
Resources: A4 paper

1. Give each student a sheet of paper and a set of decimal numbers, e.g. 2.45, 3.07, 4.6, 3.90, 2.42. Ask them to record the numbers in order from smallest to largest.
2. Have students select one of the numbers with two decimal places and write it in words.
3. Ask students to draw a number line and place all of the listed numbers. Point to the number line and ask questions, e.g. 'Why did you place 4.6 here? Would 3.4 be smaller or larger than 4.6?'
4. For students who require extension, give them a decimal number, e.g. 4.126. Ask them to round the number to the nearest hundredth and add the number to their number line (marked in a different colour).

Fractions and decimals
MA2-7NA represents, models and compares commonly used fractions and decimals

Year 4: Assessment Task Card

4.18

Decimals to Two Decimal Places — TARGETED ASSESSMENT

If the student is experiencing difficulty:

Q1 Review the ordering of numbers with no decimal places.

Q2 Use the cards from BLM 34 'Decimal Numbers and Words 1'. Give the student a number card and have them find the matching word card.

Q3 Use LO: *L2005 'Scale matters: hundreds'* and have the student review the use of number lines.

If the student has demonstrated an understanding beyond the skills, consider:

Q4 Having the student order a set of decimal numbers to three decimal places and place them on a number line.

Fractions and decimals
MA2-7NA represents, models and compares commonly used fractions and decimals

Year 4: Assessment Task Card

Chance
Resources: A4 paper

1 Give each student a sheet of paper and a list of events, e.g. will go home today, will have sport today, will eat dinner at 6 pm. Have students:

a use appropriate terminology to describe the chance of each event happening

b order the events from 'least likely' to 'most likely'.

2 Have students write two examples of everyday events where one cannot happen if the other happens.

3 Have students explain what happens if we have some counters in a bag, one counter is taken out, the colour recorded and the counter replaced.

Chance
MA2-19SP describes and compares chance events in social and experimental contexts

Year 4: Assessment Task Card

Chance TARGETED ASSESSMENT

If the student is experiencing difficulty:

Q1a Have the student practise identifying which events are likely and which are not likely. Play matching activities, using BLM 36 'Chance Words' with particular events.

Q1b Have the student practise ordering chance language and then chance events using BLM 36 'Chance Words'. This could be completed in timeline format.

Q2 Review chance situations that are everyday events where one cannot happen if the other happens. Revisit the examples created by students both in Lesson Plan 2 and as home tasks.

Q3 Review Student Book p. 78 'Five Counters in a Bag' from Lesson Plan 3. Have students repeat the activity with different numbers and combinations of counters.

Chance
MA2-19SP describes and compares chance events in social and experimental contexts

Year 4: Assessment Task Card

Collecting Data
Resources: A4 paper

1 Give each student a sheet of paper and a topic, e.g. 'Did you walk to school today?' and 'Will (or did) you watch TV today?' Have students survey their classmates and collect the data.

2 Then have students represent their data either on a Venn diagram or in a 2-way table.

3 Have students list three statements they can make as a result of their data collection.

4 For students who require extension, have them write a survey question that has only two possible answers on the topic of books.

Data
MA2-18SP selects appropriate methods to collect data, and constructs, compares, interprets and evaluates data displays, including tables, picture graphs and column graphs

Year 4: Assessment Task Card 4.20

Collecting Data TARGETED ASSESSMENT

If the student is experiencing difficulty:

Q1 Review strategies for collecting data, e.g. by lists or tables. Have the student work with small amounts of data, moving to larger sets.

Q2–3 Review a simple Venn diagram and have the student interpret the information. Continue making Venn diagrams with materials such as two hoops and sticky notes. Review the structure of a 2-way table. Have the student create a table on the floor using metre rulers, and sticky notes or masking tape. Practise reading information in the tables.

If the student has demonstrated an understanding beyond the skills, consider:

Q4 Having the student work with larger data sets using tools such as *Excel* to aid in the process of data collection and presentation.

Data
MA2-18SP selects appropriate methods to collect data, and constructs, compares, interprets and evaluates data displays, including tables, picture graphs and column graphs

Year 4: Assessment Task Card

Number Patterns
Resources: A4 paper

1 Give each student a sheet of paper and a number pattern, e.g. 3, 7, 11, 15, and have them:
a continue the pattern with the next three numbers
b write a description of the pattern.

2 Give students a starting number, e.g. 4, and have them write:
a a pattern based on addition
b a different pattern based on multiplication
c a description for each of their patterns.

3 Have students give two examples of where number patterns can be found in everyday life.

4 For students who require extension, have them write a number pattern starting at 200 and count backwards, writing a description of the pattern.

Patterns and algebra
MA2-8NA generalises properties of odd and even numbers, generates number patterns, and completes simple number sentences by calculating missing values

Year 4: Assessment Task Card

Number Patterns TARGETED ASSESSMENT

If the student is experiencing difficulty:

Q1 Revisit number patterns and how to find the pattern if provided with a number set. Give the student simple number patterns and have them identify the patterns. Have the student describe the main elements of the pattern, i.e. the starting number, whether it is addition or multiplication and what the operation is, e.g. + 2.

Q2 Review the difference between a pattern based on addition and one based on multiplication, i.e. a multiplication pattern increases much quicker.

Q3 Re-examine the number patterns on the collages students made in Lesson Plan 3, Independent Tasks, Task 1.

If the student has demonstrated an understanding beyond the skills, consider:

Q4 Having the student work with numbers both forwards and backwards and with large numbers.

Patterns and algebra
MA2-8NA generalises properties of odd and even numbers, generates number patterns, and completes simple number sentences by calculating missing values

Year 4: Assessment Task Card

Angles
Resources: A4 paper, rulers

1 Give each student a sheet of paper and ask them to draw a right angle and label it. Allow students to use the equipment as they require.
2 Have students draw an acute angle and label it.
3 Draw an obtuse angle. Have students identify it.
4 For students who require extension, have them write the numbers 4 and 5 on their page, and identify any angles in the numbers.

Angles
MA2-16MG identifies, describes, compares and classifies angles

Year 4: Assessment Task Card

4.22

Angles TARGETED ASSESSMENT

If the student is experiencing difficulty:

Q1 Revisit what a right angle is, using a protractor and in activities such as those in Lesson Plan 1, Tuning In. Identify the name of the angle with various examples and have the student draw right angles.

Q2–3 Have the student compare different angles to right angles. Have the student order angles from smallest to largest and identify which are less than or greater than a right angle and then name them.

If the student has demonstrated an understanding beyond the skills, consider:

Q4 Having the student identify different types and sizes of angles in different shapes and objects.

Angles
MA2-16MG identifies, describes, compares and classifies angles

Year 4: Assessment Task Card

Equivalent Fractions

Resources: A4 paper

1 Give each student a sheet of paper. Ask them to record the fraction $\frac{3}{4}$, and to draw a diagram showing $\frac{3}{4}$.

2 Ask students to give an equivalent fraction to $\frac{3}{4}$, and draw a diagram to represent this.

3 Say: 'You have a long chocolate bar divided into tenths, where $\frac{3}{10}$ are strawberry, $\frac{3}{10}$ are peppermint and $\frac{4}{10}$ are plain chocolate.' Have students draw a diagram and colour the chocolate bar.

4 Ask: 'What would be $\frac{4}{10}$ be equivalent to in fifths?'

5 For students who require extension, have them write three equivalent fractions for $\frac{1}{4}$.

Fractions and decimals

MA2-7NA represents, models and compares commonly used fractions and decimals

Year 4: Assessment Task Card

Equivalent Fractions

TARGETED ASSESSMENT

If the student is experiencing difficulty:

Q1 Review the process of writing fractions: What number is the numerator and what does it represent? What number is the denominator and what does it represent? Review drawing diagrams of fractions (NTO 4.16 'Comparing Shapes' could be used).

Q2–4 Review equivalent fractions (NTO 4.16 'Comparing Shapes' could be used). Revisit the 'Half and Half Again' charts from Lesson Plan 1, Tuning In, perhaps extending to quarters.

Q3 Practise dividing strips into equal amounts and then identifying the value of each section, and colouring the sections. This could be supported with the student's fraction wall from Lesson Plan 3, Independent Tasks, Task 2.

If the student has demonstrated an understanding beyond the skills, consider:

Q5 Having the student extend to more complex fractions and converting between equivalent fractions without physical models.

Fractions and decimals

MA2-7NA represents, models and compares commonly used fractions and decimals

Year 4: Assessment Task Card

Counting with Fractions

Resources: A4 paper

1 Give each student a sheet of paper and two fractions: $\frac{5}{3}$ and $1\frac{1}{6}$. Have them identify the mixed number.
2 Have students change the mixed number to an improper fraction.
3 Have students compare the two fractions, and identify which one is larger.
4 Have students start at $1\frac{1}{6}$, and write the next seven numbers in the counting sequence, counting by sixths.
5 Ask students to draw a number line showing the counting sequence.

Fractions and decimals

MA2-7NA represents, models and compares commonly used fractions and decimals

Year 4: Assessment Task Card

4.24

Counting with Fractions

TARGETED ASSESSMENT

If the student is experiencing difficulty:

Q1 Review the difference between mixed numbers and improper fractions. The student could complete sorting activities with cards.

Q2 Review the process of changing a mixed number to an improper fraction. Provide the student with simple examples and build to more complex ones.

Q3 Have the student compare fractions with common denominators, then move to denominators of the same family. The student could use models to make comparisons.

Q4 Have the student order cards in the correct counting sequence, then move to writing simple fraction counting sequences.

Q5 Have the student label missing fractions on number lines, then move to ordering fractions on number lines and, finally, creating the number lines. Work with familiar fractions first, e.g. quarters.

Fractions and decimals

MA2-7NA represents, models and compares commonly used fractions and decimals

Year 4: Assessment Task Card

Time
Resources: A4 paper, NTO 4.17 'Clocks: Advanced'

1 Give each student a sheet of paper. Use NTO 4.17 'Clocks: Advanced' and show students a time on the analogue clock. Have them record the digital equivalent. Repeat this, so students have three times recorded.
2 Have students sketch an analogue clock, and show 7:45 (a quarter to 8) on the clock.
3 Have students write three time facts, e.g. 1 minute = 60 seconds.
4 Ask students to write three activities where am and pm time is important.
5 For students who require extension, have them draw and write all of the representations they can for 3:15 pm.

Time
MA2-13MG reads and records time in one-minute intervals and converts between hours, minutes and seconds

Year 4: Assessment Task Card

Time TARGETED ASSESSMENT

If the student is experiencing difficulty:

Q1–2 Review time using NTO 4.17 'Clocks: Advanced'. Have the student record times, as well as come to the board and create times. Review the different elements of an analogue clock.

Q3 Revisit time facts and work with the student's conversion chart from Lesson Plan 2, Independent Tasks, Task 2. Revisit common links, e.g. 1 minute = 60 seconds. The student could create and play a matching game of cards with equivalent values.

Q4 Review the difference between am and pm. Discuss when am and pm are used. Ideas can be found in media such as newspapers, e.g. TV guides, weather and tide charts, advertisements.

If the student has demonstrated an understanding beyond the skills, consider:

Q5 Challenging the student with more complex times to the minute, on both digital and analogue clocks. The student could also work with 24-hour time.

Time
MA2-13MG reads and records time in one-minute intervals and converts between hours, minutes and seconds

Year 4: Assessment Task Card

Time Problems

Resources: BLM 48 'Train Timetable', A4 paper

1 Give a copy of BLM 48 'Train Timetable' and a sheet of paper to each student. Ask them to write down what format the times are given in.

2 Have students find and record the answers to:

a If I get on the train at Westville at 09:54, what time will I arrive at North Station?

b How long is the trip between Woodside and Pikefield? Give an example.

c If I had to be in Kingsville by 5 pm, what is the latest train I could catch from South Station?

d If the earliest I could leave Concord is 1 pm, what is the earliest I could be at South Station?

3 For students who require extension, have them find the time taken for the whole train trip, and the length of time between each station.

Time

MA2-13MG reads and records time in one-minute intervals and converts between hours, minutes and seconds

Year 4: Assessment Task Card

4.26

Time Problems TARGETED ASSESSMENT

If the student is experiencing difficulty:

Q1 Review the different time formats: analogue, digital, 24-hour, am and pm. Have the student complete card sorting activities. The cards could be created from BLM 44 'Clocks', BLM 46 'am and pm Times' and BLM 47 '24-Hour Times'.

Q2 Review how to read timetables. Have the student look at the time formats, how the information is organised and presented, as well as how to locate information in the timetable. Practise with a variety of transport timetables.

If the student has demonstrated an understanding beyond the skills, consider:

Q3 Challenging the student with problems, e.g. finding the lengths of journeys using more complex timetables across 24-hour periods.

Time

MA2-13MG reads and records time in one-minute intervals and converts between hours, minutes and seconds

Year 4: Assessment Task Card

Fractions and Decimals

Resources: A4 paper, three 10-sided dice, BLM 50 'Hundreds Squares'

1 Give each student a sheet of paper and three 10-sided dice. Have students roll two dice and write a decimal number with the numbers rolled, e.g. 3.2.

2 Have students write the decimal as a fraction and in words.

3 Have students roll three dice and write a decimal number with two decimal places.

4 Have students write the decimal as a fraction and in words.

5 Give students a hundreds square from BLM 50 'Hundreds Squares' and have them shade the square to represent the decimal part of the number.

6 For students who require extension, have them circle the largest of the two numbers.

Fractions and decimals

MA2-7NA represents, models and compares commonly used fractions and decimals

Year 4: Assessment Task Card

Fractions and Decimals TARGETED ASSESSMENT

If the student is experiencing difficulty:

Q1 & 3 Review what a decimal number is, with the support of a place-value chart.

Q2 Review the links between tenths as a fraction and as a decimal.

Q4 Review the links between hundredths as a fraction and as a decimal. This could be supported with NTO 4.14 'Hundred Chart: FDP'.

Q5 Using NTO 4.14 'Hundred Chart: FDP', have the student practise recognising fractions and decimals from the shaded image, and then vice-versa with the student shading the square.

If the student has demonstrated an understanding beyond the skills, consider:

Q6 Providing a larger set and a mix of decimals and fraction representations for the student to order and identify as largest and smallest.

Fractions and decimals

MA2-7NA represents, models and compares commonly used fractions and decimals

Year 4: Assessment Task Card

Unit 28 **Number Sentences**
A4 paper

1. Give each student a piece of paper and a missing number sentence (according to their ability) with addition, e.g.
   $14 + \square = 40$. Have students solve for the missing number.
2. Give students a missing number sentence (according to their ability) with subtraction, e.g. $50 - \square = 36$. Have students solve for the missing number and then write an explanation of how they solved the problem.
3. Give students a problem, e.g. 'When a number is added to 17, the answer is the same as 56 minus 21. What is the number?' Have students solve the problem.
4. For students who require extension, have them write a missing number sentence that has both addition and subtraction, and have them solve the problem.

Patterns and algebra
MA2-8NA generalises properties of odd and even numbers, generates number patterns, and completes simple number sentences by calculating missing values

Year 4: Assessment Task Card

4.28

Number Sentences — TARGETED ASSESSMENT

If the student is experiencing difficulty:

Q1–2 Have the student work with equations based on basic facts, e.g. to 10 or 20. Have the student model the equations with equipment to find the missing number.

Q3 Review problem-solving techniques, e.g. break down the problem, identify the operation, express as an equation. Allow the student to use a calculator if the numeric calculations are too difficult.

If the student has demonstrated an understanding beyond the skills, consider:

Q4 Providing examples that include multiplication and division. The student could also be encouraged to work with larger numbers.

Patterns and algebra
MA2-8NA generalises properties of odd and even numbers, generates number patterns, and completes simple number sentences by calculating missing values

Year 4: Assessment Task Card

Displaying Data

Resources: BLM 17 '1 cm Grid Paper', data set from a newspaper

1 Give each student a copy of BLM 17 '1 cm Grid Paper'. Have students devise a question about school to collect data on, e.g. favourite subject, favourite part of the day, favourite excursion.

2 Allow students a set time to collect data.

3 Have students represent their data as a graph on the grid paper.

4 Have students write three statements about their collected data.

5 For students who require extension, provide them with data from a source such as a newspaper and have them interpret. Have students write three statements about the data.

Data

MA2-18SP selects appropriate methods to collect data, and constructs, compares, interprets and evaluates data displays, including tables, picture graphs and column graphs

Year 4: Assessment Task Card

Displaying Data TARGETED ASSESSMENT

If the student is experiencing difficulty:

Q1 Give the student a range of questions to select from or key words to develop ideas.

Q2 Review efficient methods of collecting data, e.g. in a table.

Q3 Review how to construct either a picture graph or a column graph, by using objects or values on the axis.

Q4 Verbalise observations of graphs, e.g. 'Which is the greatest? Which is the least represented? Which values have equal representation?' Revisit Student Book p. 118 'Column Graphs' from Lesson Plan 3.

If the student has demonstrated an understanding beyond the skills, consider:

Q5 Providing a larger set of data that has values in the hundreds, so the student has to consider and develop the scale of the graphical representation.

Data

MA2-18SP selects appropriate methods to collect data, and constructs, compares, interprets and evaluates data displays, including tables, picture graphs and column graphs

Year 4: Assessment Task Card

Interpreting Data

Resources: BLM 53 'Fruit Graph', graph from a newspaper

1 Give each student a copy of BLM 53 'Fruit Graph'. Have them answer the questions:
   a What is the most popular fruit?
   b What two fruits have the same popularity?

2 Have students write three different facts under the graph.

3 Have students create a table of values from the graph.

4 Ask students to describe which is their preferred representation of the data, and why.

5 For students who require extension, provide them with a graph from a source such as a newspaper and have them create a table of values.

Data
MA2-18SP selects appropriate methods to collect data, and constructs, compares, interprets and evaluates data displays, including tables, picture graphs and column graphs

Year 4: Assessment Task Card

4.30

Interpreting Data

TARGETED ASSESSMENT

If the student is experiencing difficulty:

Q1 Practise interpreting simple graphs where 1 symbol = 1 data value.

Q2 Review interpreting information from graphs and tables. Work with the straightforward information first, e.g. most common, least common, items with particular values, items that are equal.

Q3 Review how to construct a table of values, e.g. labels on columns and rows, how to represent items.

Q4 Review different data representations, including picture graphs, column graphs, tables and Venn diagrams.

If the student has demonstrated an understanding beyond the skills, consider:

Q5 Providing a graph that has larger values to work with and interpret. Have the student interpret as a different type of graph, as well as a table of values.

Data
MA2-18SP selects appropriate methods to collect data, and constructs, compares, interprets and evaluates data displays, including tables, picture graphs and column graphs

Year 4: Assessment Task Card

Money

Resources: A4 paper, BLM 54 'Coins', BLM 55 'Banknotes', glue

1 Give each student a selection of notes and coins from BLM 54 'Coins' and BLM 55 'Banknotes', according to their abilities. Have students stick these to the top of a sheet of paper.

2 Have students find the total of all of the coins.

3 Have students find the change of the coins from the nearest dollar.

4 Have students find the total of all of their notes and coins.

5 Give students three prices, e.g. $2.50, $3.75 and $1.21. Have them find the total (using calculators if required) and then find the change from $10.00.

6 Ask students to list three things they discovered about money from another country.

7 For students who require extension, have them state the different coin combinations that could be used to make $4.85.

Addition and subtraction (money)

MA2-5NA uses mental and written strategies for addition and subtraction involving two-, three-, four- and five-digit numbers

Fractions and decimals (money)

MA2-7NA represents, models and compares commonly used fractions and decimals

Year 4: Assessment Task Card

Money TARGETED ASSESSMENT

If the student is experiencing difficulty:

Q2 & 4 Have the student practise counting techniques. Remind the student of strategies, e.g. starting with the largest value and counting on.

Q3 & 5 Review what change is. Again review strategies, e.g. counting up to. Use real money or coins and notes from BLM 54 'Coins' and BLM 55 'Banknotes'.

Q5 Have the student review addition. Use calculators, and check that the student is putting in the decimal values and reading the answers correctly.

Q6 Review money from different countries, particularly one with a LOTE focus. NTO 4.18 'Money from Other Countries' could also be used. Revisit the posters developed by the student in Lesson Plan 3, Independent Tasks, Task 1.

If the student has demonstrated an understanding beyond the skills, consider:

Q7 Providing the student with longer lists of items to find the total of, with more rounding elements, and also to find the change.

Addition and subtraction (money)

MA2-5NA uses mental and written strategies for addition and subtraction involving two-, three-, four- and five-digit numbers

Fractions and decimals (money)

MA2-7NA represents, models and compares commonly used fractions and decimals

Year 4: Assessment Task Card

Word Problems
Resources: A4 paper

1 Give each student a sheet of paper. Read out a word problem. For example: 'At the bus station there were 10 buses, which had 8 wheels each. How many wheels were there?' Have students draw a diagram of the problem.
2 Have students write and solve a related number sentence.
3 Have students write a brief explanation of how they solved the problem.
4 Give students a division problem to solve involving a remainder, e.g. 77 ÷ 9 =.
5 Have students then write a word problem for the problem in Question 4.
6 For students who require extension, have them write a word problem that will require the person solving the equation to use both addition and subtraction.

Multiplication and division
MA2-6NA uses mental and informal written strategies for multiplication and division

Year 4: Assessment Task Card

Word Problems TARGETED ASSESSMENT

If the student is experiencing difficulty:

Q1–3 Give the word problem in written, rather than verbal form. If the student cannot draw a diagram of the equation, provide some modelling equipment, e.g. counters, to represent the wheels.

Q4 Review strategies to solve division problems with or without remainders (see Unit 16, Lesson Plan 3, p. 82).

Q5 Provide modelling equipment, then have the student model the equation, and then try to write a word problem. The student could also work with smaller numbers.

If the student has demonstrated an understanding beyond the skills, consider:

Q6 Having the student write and/or solve problems that require more than one step. For example: 'At the shop I bought 3 items that cost $2.50 each. What was the total change I received from $10.00?'

Multiplication and division
MA2-6NA uses mental and informal written strategies for multiplication and division

Year 4: Assessment Task Card

Patterns

Resources: a pattern with symmetry, BLM 17 '1 cm Grid Paper', coloured pencils or felt pens

1 Display a pattern containing symmetry, either on the IWB or on hard copy (e.g. in a book).
Have students identify:
a the shapes in the pattern
b if the pattern has symmetry and how they know.

2 Give each student a copy of BLM 17 '1 cm Grid Paper', and ask them to create a simple pattern with an element of symmetry.

3 Have students describe their pattern.

Two-dimensional space

MA2-15MG manipulates, identifies and sketches two-dimensional shapes, including special quadrilaterals, and describes their features

Year 4: Assessment Task Card

Patterns TARGETED ASSESSMENT

***If the student is experiencing difficulty*:**

Q1a Review different common 2D shapes. Practise identifying each of the shapes by name (matching activities could aid in this area).

Q1b Review the concept of symmetry. Symmetry in nature could be examined, e.g. butterflies, flowers and faces.

Q2–3 Have the student practise creating patterns with concrete materials, e.g. pattern blocks, or electronically. Extend their understanding by incorporating symmetry into the patterns.

Two-dimensional space

MA2-15MG manipulates, identifies and sketches two-dimensional shapes, including special quadrilaterals, and describes their features

Test A: Student Sheet

Name: ______________________ Date: ______________

Whole numbers

1 Circle the answer.

When two **odd** numbers are added together the result is:

odd  neither

2 State if the answer to 6 × 4 is odd or even.

Even (6 × 4 = 24)

3 Write twenty-six thousand, five hundred and seven as a numeral.

26 507

4 Order the numbers from **smallest** to **largest**.

32 461 32 872 32 048 31 796

31 796, 32 048, 32 461, 32 872

5 Write 52 560 in expanded form.

50 000 + 2 000 + 500 + 60

6 What is the value of the underlined digit?

4<u>7</u> 908 7 000

Patterns and algebra

7 Continue the number sequence.

24, 30, 36, 42, 48, 54

Test A: Student Sheet

8 Write the number sequence that begins at 40 and counts forward by 4 for the next 5 terms.

40, 44, 48, 52, 56, 60

9 What is the pattern in the following number sequence?

90, 81, 72, 63, 54, ...

Begin at 90, subtract 9.

Multiplication and division

10 If 7 × 5 = 35 then 35 ÷ 7 = 5

11 Write a related multiplication equation for:

3 × 4 = 12 or

4 × 3 = 12

12 True or false?

4 × 7 = 7 × 4 True

13 Multiply the numbers shown on the dice.

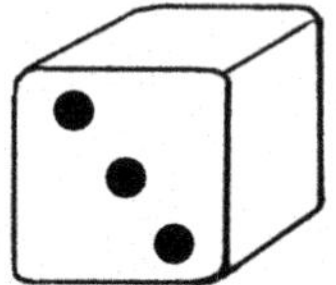

15

Multiply the answer by 10. 150

Test A: Student Sheet

Length and temperature

14 What measurement is indicated on the ruler?

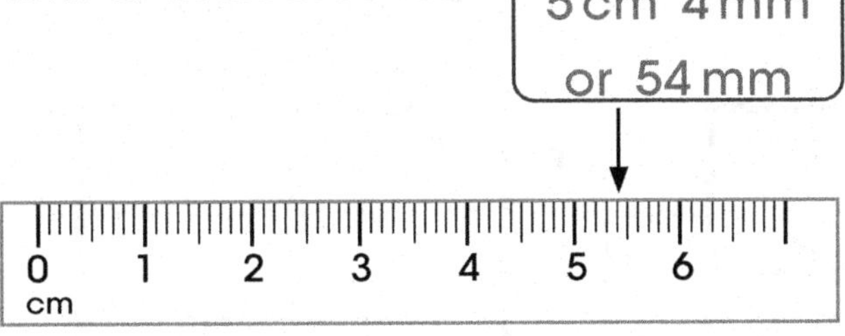

Volume and capacity

15 Shade jug B so that it shows half a litre.

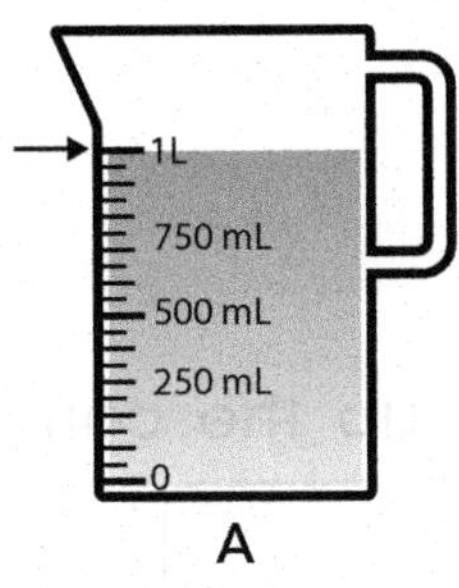

A

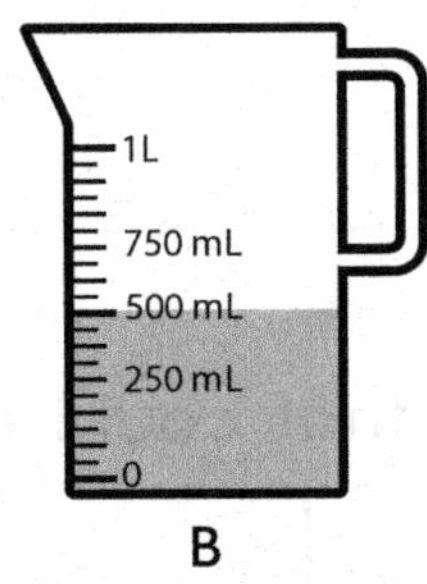

B

16 Colour the thermometer to show 10.5°C.

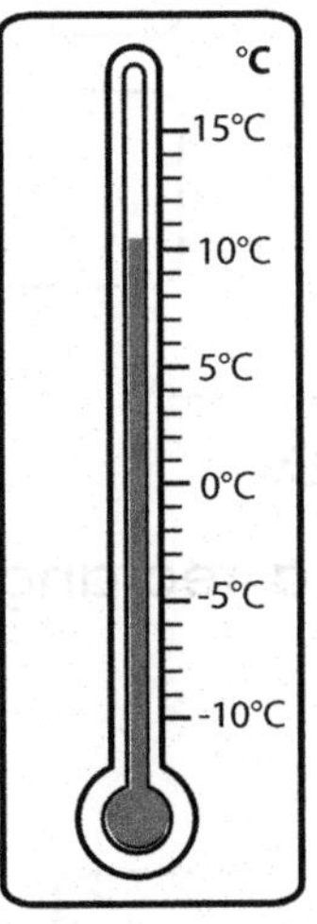

17 The box holds cubes of 1 cubic centimetre.
What is the volume of the box?

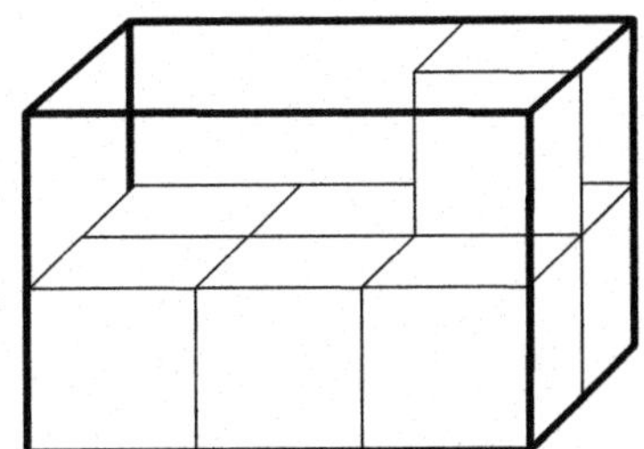

$12\,cm^3$

Test A: Student Sheet

4.A

Area

18 Which object has the **largest** area – A or B?

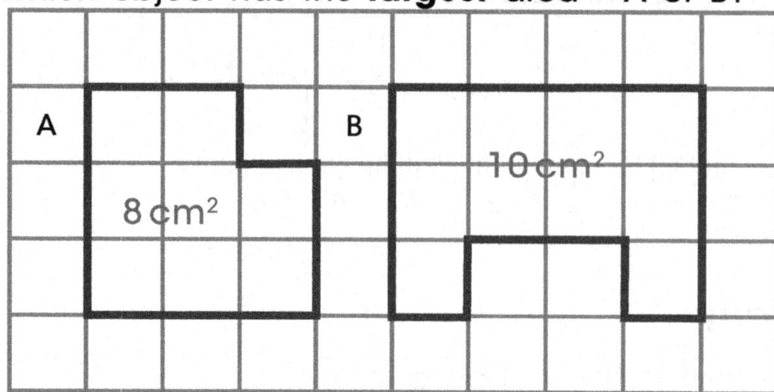

B

Two-dimensional space

19 Name the two shapes that could make up the composite shape.

Rectangle

Triangle

Three-dimensional space

20 Finish this drawing of a rectangular prism.

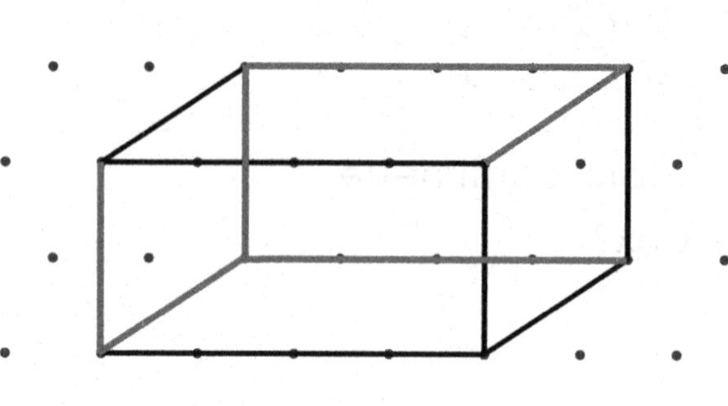

Test A: Student Sheet

Position

21 Tom was at the art gallery. He followed the path passing through the café to the location in the top right corner of the map. Where was Tom?

Park

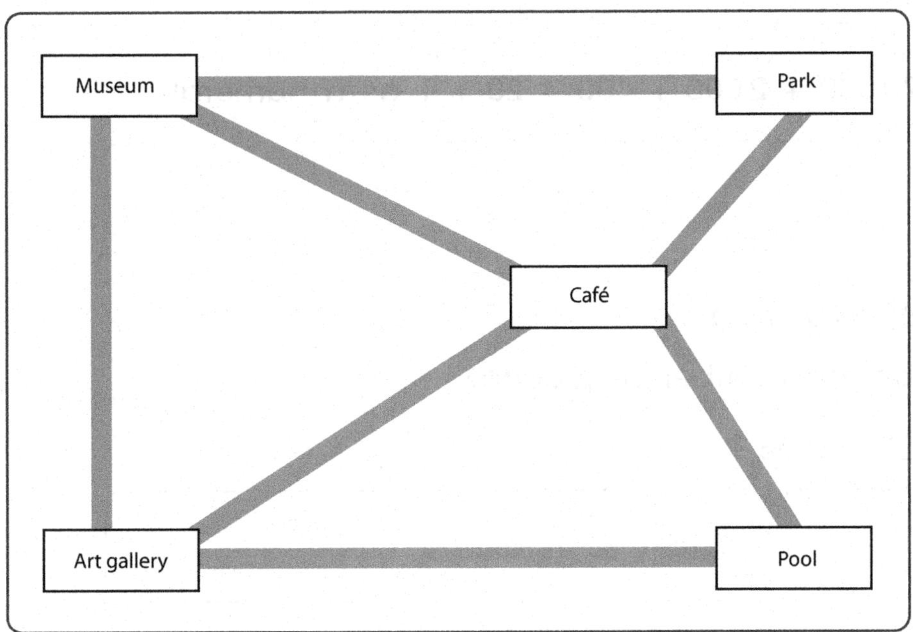

22 Look at the different-shaped islands. How far is it from the circle island to the triangle island if one square length represents 10 metres?

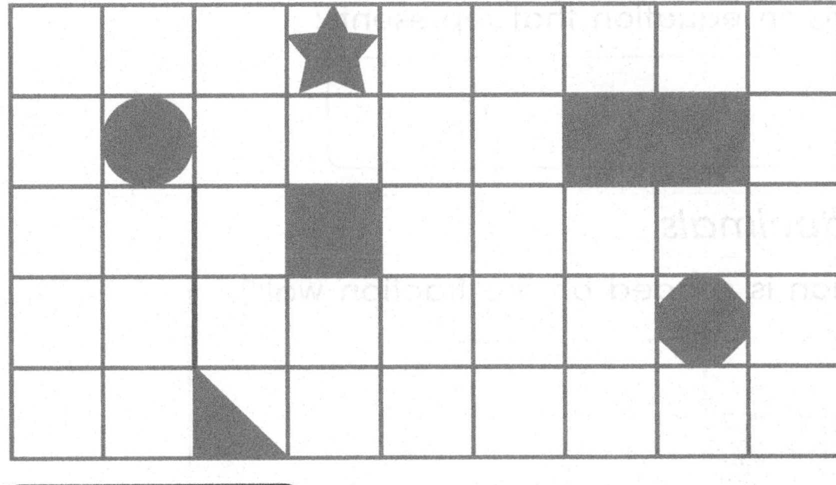

20 metres

Test B: Student Sheet

4.B

Name: ______________________ Date: ______________

Whole numbers

1 Write 42 030 in words.

Forty-two thousand and thirty

2 Write 70 000 + 2000 + 400 + 20 + 1 as a numeral.

72 421

Patterns and algebra

3 Describe the number sequence.

7, 14, 21, 28, 35

Start at 7, add 7.

Multiplication and division

4 Write the correct numbers in the boxes.

If 8 × 9 = [72]   Then [72] ÷ 8 = 9

5 Write a division equation that represents:

half of 20.   20 ÷ 2 = 10

Fractions and decimals

6 What fraction is shaded on the fraction wall?

$\frac{1}{4}$

Test B: Student Sheet 4.B

7 State 2 as an improper fraction.

$\frac{2}{1}$

8 Label the box on the number line.

$\frac{3}{4}$

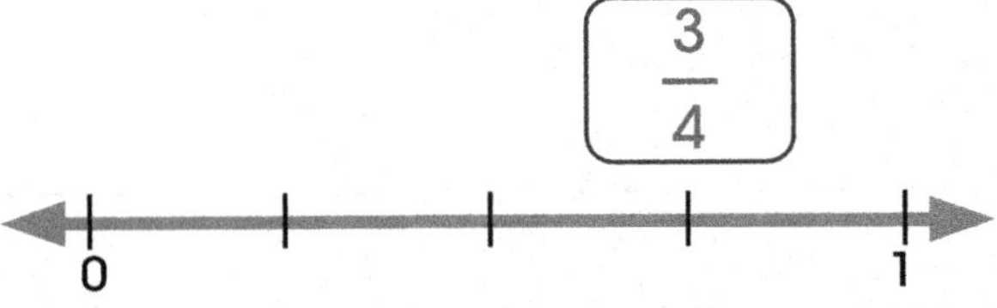

9 Circle the largest value.

(1.10) 0.10 0.01

10 4.6 written as a fraction is:

$4\frac{6}{10}$

Addition and subtraction

11 Find the total change from $10 if the book and pen are purchased.

$1.80

Patterns and algebra

12 Describe the two number patterns found on an analogue clock face.

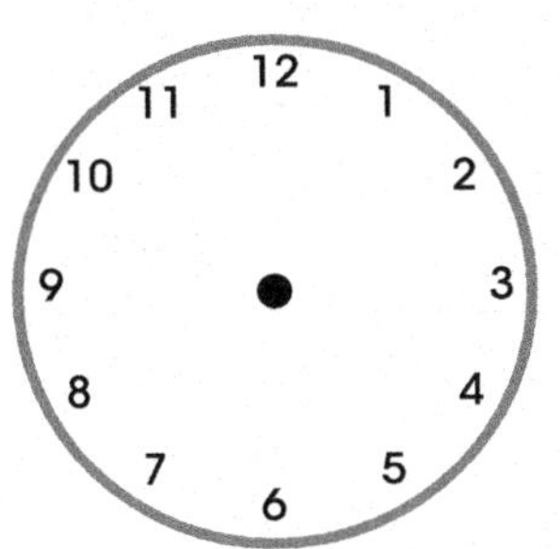

hours: ones

minutes: fives

13 There were 10 trolleys at the supermarket, each with 4 wheels. How many wheels were there?

Write a number sentence and solve it.

$10 \times 4 = 40$

14 Write the correct number in the box.

$4 \times$ [6] $= 20 + 4$

Volume and capacity

15 What volume is represented by the model made with interlocking centimetre cubes?

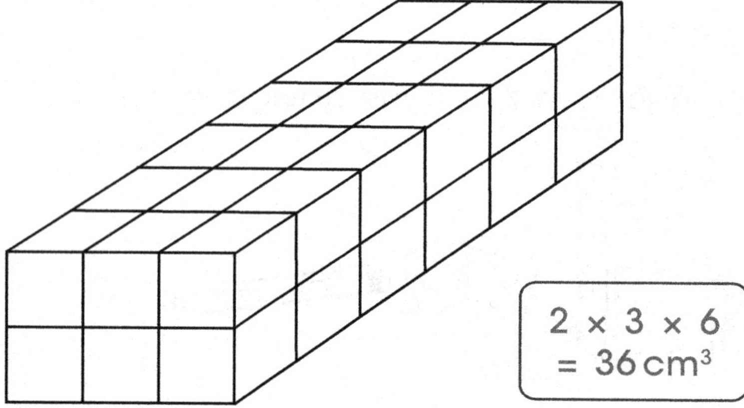

$2 \times 3 \times 6 = 36\,cm^3$

Time

16 It is two hours until lunchtime. Which equation is the correct one to find the total number of minutes?

A 2×90

B 2×60

C $60 + 60 + 60$

D 2×24

B

Test B: Student Sheet

If I caught the bus from school and got off at the beach, how long was my trip?

School	Shops	Beach
4:00	4:15	4:25
4:15	4:30	4:40
4:30	4:45	4:55

25 minutes

Angles

Is the following angle greater than or less than a right angle?

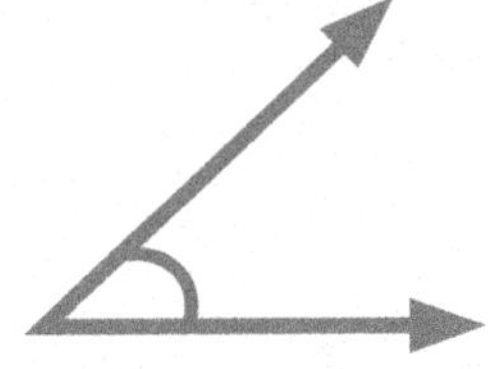

less than

Two-dimensional space

Is the pattern symmetrical? Answer "Yes" or "No".

Yes

Chance

Order the events from **least likely** to **most likely**.

A The sun will rise.

B A whale will visit my school today.

C It will be windy today.

B, C, A

Lucy had a baby girl. The chance her next baby will be a boy is:

50% or equal chance

Test B: Student Sheet

4.B

Data

22 Use the data in the graph to complete the table.

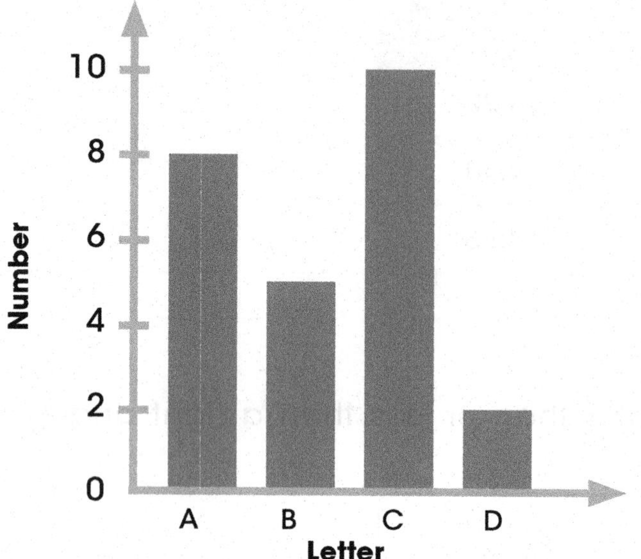

Letter	Number
A	8
B	5
C	10
D	2

23 The number of cloudy days in July was:

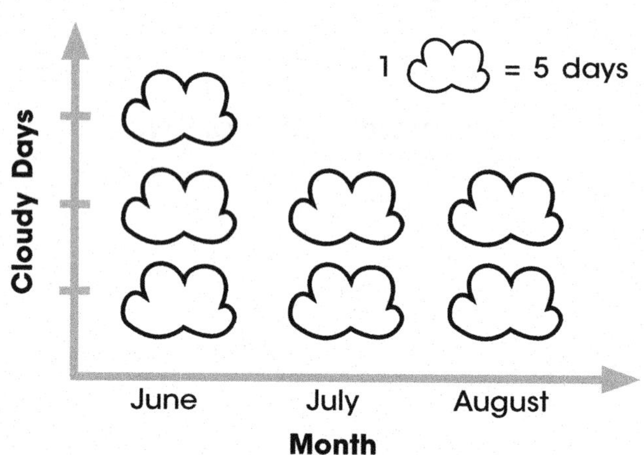

10 days

24 The **least** popular colour ball was:

Number of balls

100

80

60

40

20

0

Red Blue Green Yellow

Colour

Green

Class Unit Overview

Class: __________ **Year:** ______

33																								
32																								
31																								
30																								
29																								
28																								
27																								
26																								
25																								
24																								
23																								
22																								
21																								
20																								
19																								
18																								
17																								
16																								
15																								
14																								
13																								
12																								
11																								
10																								
9																								
8																								
7																								
6																								
5																								
4																								
3																								
2																								
1																								
Initials																								

Class SB Assessment Overview Class: ________ Year: ______

Initials	1/p.7	2/p. 11	3/p. 15	4/p. 19	5/p. 23	6/p. 27	7/p. 31	8/p. 35	9/p. 39	10/p. 43	11/p. 47	12/p. 51	13/p. 55	14/p. 59	15/p. 63	16/p. 67	17/p. 71	18/p. 75	19/p. 79	20/p. 83	21/p. 87	22/p. 91	23/p. 95	24/p. 99	25/p. 103	26/p. 107	27/p. 111	28/p. 115	29/p. 119	30/p. 123	31/p. 127	32/p. 131	33/p. 135

Class Assessment Task Card Overview

Class: __________ Year: ______

Initials	4.1	4.2	4.3	4.4	4.5	4.6	4.7	4.8	4.9	4.10	4.11	4.12	4.13	4.14	4.15	4.16	4.17	4.18	4.19	4.20	4.21	4.22	4.23	4.24	4.25	4.26	4.27	4.28	4.29	4.30	4.31	4.32	4.33

WEEKLY MATHS PLANNER

Unit: ______ Week: ______ Term: ______ Date: ______ Year Level: ______

Resources: ____

		Monday	Tuesday	Wednesday	Thursday	Friday
	Tuning In					
	Whole-Class Introduction					
Small Group Focus	Independent Tasks (individual, pair, small group)					
Small Group Focus	Teaching Group					
	Reflection					

TERM MATHS PLANNER

Year Level: ______________ Term: ______________ Year: ______________

Week	Unit/page no.	Strand/Sub-strand/Content description/Code

YEARLY MATHS PLANNER

Year: ________________ Year Level: ________________

Term 1	Unit
Week 1	
Week 2	
Week 3	
Week 4	
Week 5	
Week 6	
Week 7	
Week 8	
Week 9	
Week 10	

Term 2	Unit
Week 1	
Week 2	
Week 3	
Week 4	
Week 5	
Week 6	
Week 7	
Week 8	
Week 9	
Week 10	

Term 3	Unit
Week 1	
Week 2	
Week 3	
Week 4	
Week 5	
Week 6	
Week 7	
Week 8	
Week 9	
Week 10	

Term 4	Unit
Week 1	
Week 2	
Week 3	
Week 4	
Week 5	
Week 6	
Week 7	
Week 8	
Week 9	
Week 10	